TECHNICAL WRITING
FOR PRIVATE INDUSTRY
The A-to-Z of O&M Manuals

TECHNICAL WRITING
FOR PRIVATE INDUSTRY
The A-to-Z of O&M Manuals

Edward von Koenigseck
James N. Irvin, CPL
Sharon C. Irvin

KRIEGER PUBLISHING COMPANY
Malabar, Florida
1991

Original Edition 1991

Printed and Published by
KRIEGER PUBLISHING COMPANY
KRIEGER DRIVE
MALABAR, FLORIDA 32950

Library of Congress Cataloging in Publication Data

Koenigseck, Edward von.
    Technical writing for private
  industry / by Edward von Koenigseck, James N. Irvin, and Sharon C.
  Irvin.
      p.   cm.
    ISBN 0-89464-363-0 (alk. paper)
    1. Technical writing. I. Irvin, James N. II. Irvin, Sharon C.
  III. Title.
  T11.K64 1990
  808'.0666—dc20   89-31884
                                                       CIP

10 9 8 7 6 5 4 3 2

CONTENTS

Part I

THE PROCESS OF CREATING THE MANUAL

Part II

THE CHAPTERS IN THE MANUAL

Part III

THE SUBCONTRACTING PROCESS AND
COMPUTER-AIDED ACQUISITION AND LOGISTICS SUPPORT

LIST OF ILLUSTRATIONS

LIST OF TABLES

Foreword

The ''A-to-Z of O&M Manual-Technical Writing for Private Industry'' contains three parts. Part I describes the process of creating the manual; Part II describes the typical chapters and drawings in the manual; Part III describes the subcontracting process and data-based technical manuals. A glossary is included in Appendix A.

The following is a list of U.S. Government Specifications and Standards cited in the book: MIL-M-63036; MIL-HDBK 63038-1 (TM); MIL-HDBK 63038-2 (TM); MIL-M-38784; MIL-M-7298; MIL-M-38798. They can be obtained by writing to Superintendent of Documents, U.S. Government Printing Office, Washington, DC 20402 or Standardization Document Order Desk, 700 Robbins Avenue, Building No. 4, Section D, Philadelphia, PA 19111-5094.

This book evolved out of a need for a textbook for a course in Advanced Technical Writing. Its intent is to expose people to the process of government procurement and the contents of a typical manual and to introduce them to technical terms and military specifications. This book makes extensive use of terms, phrases, abbreviations, and acronyms common to technical publications.

Introduction

What Are Technical Writers?

Who and what are technical writers? People do not associate the results of these particular writers' efforts with their sources. A person who purchases a new car receives an Owner's Manual. The same is true for any number of household mechanical or electrical devices such as lawn mowers, washing machines, blenders, microwave ovens, compact disk players, VCR's, etc. Only the product manufacturer's name is known: the true author or authors are unknown.

The term "writer" can include numerous subclassifications. Everyone knows what a writer is. Usually the association automatically brings to mind an author or novelist, but technical writers are obscure writers. They may write any of a wide variety of documents including brochures, manufacturer specification sheets, owner's manuals, or complex technical manuals.

The quality of these "user's manuals" varies from excellent to atrocious. There are several reasons for this diversity of quality in user manuals, including the intent of the manual, the writer's ability, the funds available to write the manual, the scheduled time available to write and produce the manual, and the quality and quantity of technical information available for the writer to meet the requirements which define format and content.

This book is written so the reader will understand the task of the technical writer who prepares data for the United States Government/Military in response to a contract bid. Due to the intensely regimented nature of data prepared for the military, a writer who is able to digest the guiding documents for data preparation and who can fulfill the requirements and meet all the demands cited by these documents can also prepare material for just about any other kind of technical writing requirement.

Where Do Technical Writers Come From?

Most hardware technical writers have a background in electronics and computer science. In the past, many writers had military experience and specialized training in one or more areas of electronics, usually at the technical level. Military training may have included everything from operation to depot level maintenance. Today more writers are graduating from colleges and universities with degrees in Technical Communication.

It should be understood that just as there are many types of writers, there are also many concepts of what constitutes a technical writer. In most companies, technical writers are divided into different classes or levels based on experience and skills. One may be a junior writer, a senior writer, a project director, or a task leader or any number of classifications of writers. Then again, a senior technical writer in one company may be given a different title doing the same level of work in another company. This book will attempt to explain the diverse responsibilities of technical writers, including the management of writing tasks.

Part I

THE PROCESS OF CREATING THE MANUAL

Chapter 1

The Functions of a Technical Writer

Introduction

The prime function of a technical writer is to write technically oriented material in a clear and concise manner. This function is only one of several, because the total number of functions depends upon the requirements of the job. Table 1-1 lists the functions technical writers may be expected to perform.

The "Super" Technical Writer

After reading the list of functions a technical writer may be called upon to perform, two factors become evident: a competent writer must have a well-rounded education and must have experience. The following paragraphs describe background requirements for a technical writing career.

Electronics

The ability to "read" wiring, schematic, logic, block, and flow diagrams is as vital to a technical writer in the electronics industry as the ability to read charts is to a navigator on a ship. Unlike engineers who specialize in particular areas of design, the technical writer should be able to write data on any and all areas of electronics.

Mechanics

Electronic components are mounted on assemblies that are placed in chassis, which are mounted in racks that may be interconnected within a building or trailer or some other form of housing. The technical writer must be able to read assembly drawings, parts lists, cabling diagrams, installation drawings, and many other documents related to the hardware about which the technical manual is being written. The symbols used in these various drawings, which the writer must be able to interpret, number in the thousands. The U.S. Government has dozens of specifications and standards that identify specific symbols to be used for specific drawing types. The commercial industry has set up its own standards through various agencies such as the Institute of Electrical and Electronics Engineers (IEEE).

Graphics

A key area of any technical publication is the artwork which supports or supplements the text. The competent technical writer must be able to interpret drawings and create new drawings from mechanical and electrical engineering drawings. Most technical manuals contain some form of mechanical assembly drawings done in perspective, exploded view, front and rear view, or isometric view. These drawings are prepared by the technical writer in rough draft form or as computer-generated art. The illustrations may be submitted to an artist who creates the final art form. The writer must be familiar with contractual requirements that define the types of illustrations which are permissible. In some instances photographs may be used in conjunction with, or in place of, line drawings. The technical writer must understand how to create new illustrations using available engineering drawings and parts lists.

Production Control

Technical writers may be called upon to do more than write text and prepare rough-draft illustrations. They may find it necessary to follow through the entire process of the production sequence, from rough type formatting to preparation of the reproducible copy. They should be familiar with both the capabilities and limitations of their Production Department's capital equipment and personnel and the contractual requirements of the deliverable documents purchased by the customer.

Writing Ability

The ability to write implies that a technical writer not only knows how to present material but also understands the basic rules of grammar. The need for technical competence is as great as the need for communication skills. The more capable writer produces more accurate and usable manuscripts. Today greater stress is placed on the writer's ability to write to the desired reader's level of comprehension. In many instances material must be prepared to the specific level of the reader and described in the detail necessary to help the reader perform the required operation and maintenance tasks. The writer's task is now easier because of the availability of computer software that automatically grades the reading level of the technical material, checks spelling and provides a thesaurus.

Editing and Proofreading

In some instances technical writers may not have an editor available and will have to edit and proofread their own work. The editor's task can become quite complex since format, layout, and content are involved in final manuscript prepa-

Table 1-1 *Technical Writing Functions*

Function	Description
Writers	plan and prepare text and illustrations to convey technical data based on contractual requirements.
Illustrators	prepare final, camera-ready illustrations to support the text.
Editors/Proofreaders	edit/proofread the manuscript and artwork.
Technicians	participate in ''hands-on'' validation and verification of procedural data contained in the manual by performing operation and maintenance procedures on hardware using test equipment and tools as required.
Supervisors	assign and direct work of supporting writers who are to assist in completion of the writing task.
Word Processors*	enter data into computer for storage and recall.
Accountants	perform task cost control function through periodic reporting of expenditures versus budget.
Planners	identify personnel requirements and job status.
Liaison	meet with various engineers/scientists/ designers to obtain data required for job.
Customer Relations	meet with customers to identify their needs and to satisfy the contractually required deliverables.
Proposal Writers/Negotiators	prepare and defend bids and proposals for obtaining new business.
Subcontract Administrators	write statement of work, negotiate and administer subcontracts, define contractual requirements and provide guidance to subcontractors.
Production Coordinators	give direction to word processors, printers, coordinator, photographers, and artists.

*Note: Unless otherwise indicated, the term ''word processor'' refers to those who enter data into a computer, and not to the computer doing the word processing.

ration. The editor/proofreader should also be familiar with standard editorial symbols to indicate corrections that must be made before final printing.

Management

As technical writers expand their abilities, they are given more responsibility in the form of management tasks. A senior level writer should be able to supervise other writers, control the cost and schedule of a project, and may be involved in the company's prospective new programs through bid and proposal activities. They should be able to identify and plan milestones for a project and see that appropriate personnel are assigned to complete a project.

Psychology

Technical writers should have an understanding of basic psychology. A writer works in an environment that interacts with supervisors, other writers, engineers, and customers. Writers are often either motivating, negotiating, or engaging in information transfer with others to obtain source material. How well they succeed can be (and usually is) based on how well they can work with other people.

Contracts/Subcontracts

If part of the technical writer's task includes bid and proposal efforts and subcontract administration, writers must have a working knowledge of the types of contracts the company engages in, the company policy related to contracts and subcontracts, and the responsibilities of being a prime contractor in relation to both customers and subcontractors. A writer may not be directly involved in contract or subcontract preparation but may be required to perform a task defined by a contract. Familiarity with general contract structure and content is critical.

Integrated Logistics Support

Technical writing is one aspect of technical documentation, and documentation is a key element of Integrated Logistics Support (ILS). ILS includes test and support equipment, supply support, personnel and training, technical data, facilities, transportation and handling, and maintenance planning. ILS is generally considered to be a management function that provides initial planning, funding, and control. It assures the customers they will receive a system that will meet performance requirements and be economically and readily supportable through its designed life cycle.

Technical writers working on government programs will usually become involved with several of the ILS functions. They should understand the requirements and interrelationships of each element and the outputs from each which are applicable to the technical manual and which serve as source data.

Security Clearance

The U.S. Government has several levels of security classifications that may involve a writer. The most commonly known levels are Confidential, Secret, and Top Secret. A company which decides to do classified work for the government or military must prove that it can meet the security requirements needed to protect information related to the work being done. Assuming these requirements can be met, the company becomes a ''cleared'' facility. The next step is to obtain proper clearances for the people who will be involved in designing, building, and documenting the hardware and/or software.

Technical writers developing manuals for classified equipment may also have to obtain security clearances. Depending upon the level of clearance, several months or even a year or more may pass before the entire background search of a writer is completed.

Conclusion

Technical writers, then, perform many different functions depending on their job descriptions and places of employment, but wherever they work they must use their time wisely. Time management is a daily activity; writers must continually identify areas of their work which consume their time and are nonproductive. "Make work," which is essentially unproductive time spent, is a great destroyer of budgets. It is work not related to their tasks that can deprive them of time and funds they may need later in a program or project. By keeping a daily journal, technical writers can find out how much of their time is spent on non-work related activities and make adjustments where needed.

Chapter 2

Integrated Logistics Support—Where Technical Documents Fit In

Introduction

Both government and industry organizational structures are complex and constantly changing. There is a growing effort, however, to streamline the acquisition process. This chapter describes three overlapping cycles: the government acquisition cycle, the product life cycle, and the design/development cycle. Typical government acquisitions have five phases, the life cycle of most products and systems has six phases, and design and development of products have four phases. This chapter looks at life cycle engineering or Integrated Logistics Support (ILS) and at the role of the writer in the complex set of life cycle phases. Both the government and government vendors include technical publications under ILS.

Origin of the Term "Logistics"

When Roman legions marched to battle, the LOGISTA was the officer responsible for everything except tactics and command. He designed the engines of war, the bridges, the roads, the ramps, and the fortifications and provided food, beverages, weapons, ammunition, uniforms, armor, mounts and pack animals, housing, latrines, slaves, and tools. He planned and directed training and, if the task was neither command nor tactics, it fell under the scope of the Logista. The Logista provided every type of support function for the legionnaire in the field, whether in garrison or battle. The Logista was the first military engineer and the prototype logistician. This, then, is the origin of the term "Logistics."

Technical Publications

Within the Department of Defense (DOD), Technical Publications are but one part of Technical Data, which, in turn, is one element of Logistics Support. Publications (technical manuals) are designed to meet the individual and unique needs of each branch of the DOD (Army, Air Force, Navy and Marines). Therefore, technical data includes hundreds of document types. A very brief listing would include, but not be limited to, the following:

- Operation and Maintenance (O&M) Manuals
- Technical Manuals (TM for Army and Navy)
- Technical Orders (TO for Air Force)
- Depot Maintenance Work Manuals (DMWR for Army)
- Installation Manuals
- Modification Work Orders (MWO)
- Time Compliance Technical Orders (TCTO)
- Overhaul Manuals
- Preventive Maintenance Work Cards (PMWC)
- Function Oriented Maintenance Manuals (FOMM)

Each branch of the DoD has identified the format and content of each type of technical publication to fit its own needs.

Government Procurement of Technical Data

Government procurement of technical data is permitted under the Federal Acquisition Regulations (FAR), previously known as Defense Acquisition Regulations (DAR), and Armed Services Procurement Regulations (ASPR). These regulations define and describe the types of data and guiding principles to be followed when the government (the customer) decides to acquire data from a manufacturer (the contractor). According to ASPR 9-02 (a):

> Technical data is expensive to prepare in the required form and to maintain and update. Every effort, therefore, should be made to avoid placing a requirement upon a contractor to prepare and deliver data unless the need is positively determined.

The following quote is taken from the Integrated Logistic Support Primer, U.S. Army DARCOM Material Readiness Support Activity, p. 7:

> A material concept investigation may be initiated by either the combat developer or the material developer in response to operational concepts, material proposals, ideas, or suggestions received from any source. The impact of these technology improvements on the Army's logistic system and on operations and maintenance concepts is critical to the operational readiness of Army material.

When the government acquires hardware then, technical publications are normally included. Technical manuals are necessary for the customer to operate and maintain a piece of equipment. The size of any publication can vary from a few sheets of paper to one or more volumes. The complexity of the hardware and software, the specifications which identify manual format and content, the maintenance concept, and the user's needs determine the total number of pages of any manual.

ILS Specialists

Today logistics includes all the support needs of "the system" (whether it be one piece or a roomful of equipment): spares, tools, test equipment, packaging, preservation, consumables such as fuel and lubricants, training equipment, and manuals. The ILS team plans and designs the support system which assures the long life of the equipment and includes the following members:

● ILS Managers (ILSM)

Provide overall guidance and control of all logistic functions for the program. The ILSM must be familiar with each of the logistics specialities and how they interact with program objectives and team members.

● Logistics Engineers

Analyze the logistics support needs and document those needs using the Logistics Support Analysis Record (LSAR), do trade studies and define the most cost-effective support system alternatives, define cost-effective sets of spare parts, recommend efficient support equipment and test equipment, and define the skills and qualifications and quantity of the users, operators, and maintenance personnel.

● Configuration Management Specialists

Ensure that drawings, equipment, computer programs, and the status of their revisions are known and in proper agreement.

● Data Management Specialists

Assist the program manager and ensure timely submission of contract-required data items in the defined quantity and format.

● Provisioning Specialists

Define the provisioning (spares, supplies, etc.) recommendations as well as the packaging, preservation, handling, and transportation needs.

● Training Specialists

Define the training needs and training equipment, the number of classes, prerequisite knowledge, etc. These specialists may also plan, develop, and teach many of the classes.

● Publications Specialists

Include technical writers, editors, illustrators, and reproduction specialists who plan and produce the needed system, hardware, and software manuals. They may also assist in preparing review illustrations, data items, etc.

A Typical ILS Organization

In many companies, each of the logistics functions is a separate entity; however, as organizations grow and as the government emphasis on controlling the lifetime costs of the system it buys grows, ILS organizations become interdependent entities. Figure 2-1 shows one effective ILS organization (functional). Each of the functions previously described is represented. They include systems effectiveness, support system definition, spare parts provisioning, maintenance planning, manuals, training, packing preservation, handling, and transportation. Also included are support functions such as clerical help, quality control, and data processing support. Many of the ILS functions include

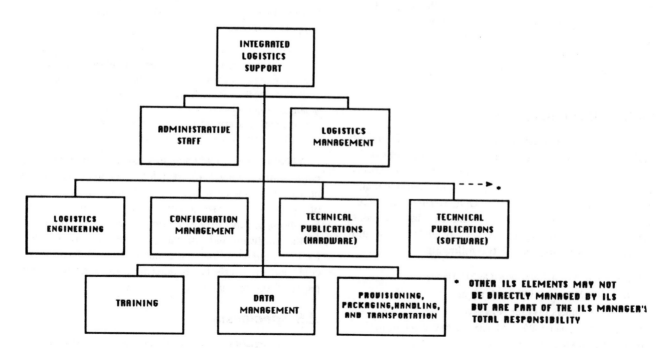

Figure 2-1 A Functional ILS Management Organization

computer simulation, computer modeling, and computerized data retention and production.

The ILS functions are critical to implementing a cost-effective system. Through each phase of a typical government acquisition, some, if not all, the ILS specialties contribute vital support to attaining the program goals.

The Acquisition Cycle

Figure 2-2 is based on a figure in MIL-STD-1388-1A (Logistic Support Analysis). It identifies the acquisition phases and shows the logistics goals and functions. The five phases of a typical government acquisition process are:

- Pre-concept
- Concept Exploration
- Demonstration and Validation
- Full Scale Development
- Production/Deployment/Post-Production

A brief glance at Figure 2-2 shows the intricately integrated logistics function. Four key government evaluations during

development and early production are defined in the following subparagraphs:

DT/OT I a Design Test and Operability Test done on the system proof item, such as a brassboard (prototype) system, a proof of concept and theory which may have little resemblance to the final system configuration.

DT/OT II another Design Test/Operability Test, but done with a Full Scale Production (FSP) version of the system which is nearly in the final fieldable configuration.

FOT&E a Field Operability Test and Evaluation using the final system configuration.

FIELD DATA Information collected in the field and used by the writers to improve and correct the manuals.

ILS in the Life Cycle

While the acquisition cycle has five phases, the life cycle of a "system" has six phases:

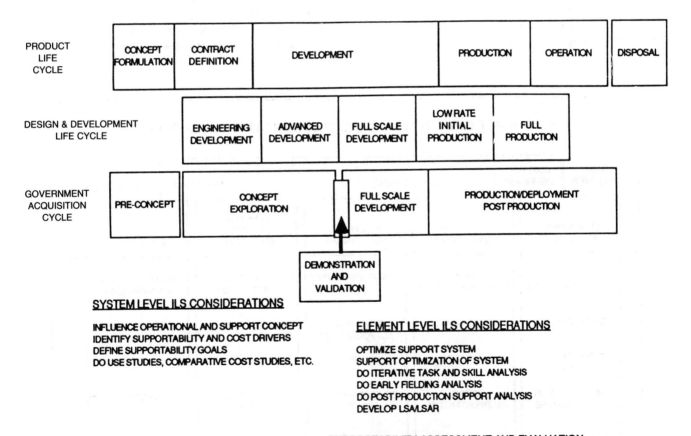

Figure 2-2 *ILS in the Five Phases of the Acquisition Life Cycle*

- Concept Formulation
- Contract Definition
- Development
- Production
- Operation
- Disposal

Industry in general, and technical writers in particular, rarely take part in the acquisition cycle except in the development and production phases. These two phases are often divided into five phases:

- Engineering Development
- Advanced Development
- Full-Scale Development (FSD)
- Low-Rate Initial Production (LRIP)
- Full Production

Figure 2-3 shows some of the logistics functions which apply in the various phases of the typical development cycle. Changes to the design in the early development phases have relatively little impact on the life-cycle cost, but during production and post production phases, the cost to implement a change is very high. Figure 2-4 illustrates the "moment of truth" relative to cost impact of change through the life cycle.

During the **Engineering Development Phase**, the designers prove that the system and the concept will work by using engineering "breadboard" circuits and computer simulations. Historically, logistics considerations such as testability, supportability, reliability, etc., have had little or no priority in the engineering realm. This concept is changing now; more stress is being placed on applying ILS functions earlier in the program. Concurrent engineering (CE) changes the traditional engineering process, integrating other disciplines such as producibility, reliability, testability, maintainability, supportability, and manufacturability with design. This significantly shortens the acquisition cycle, allows earlier system use, reduces the risk of failure, and changes the writer's role. The CE process ensures that production and support are given equal consideration as cost, schedule, and performance. Well-run programs begin to incorporate the logistics-related functions in the advanced development phase, and often the logistician must educate the designers, that is show them the value of built-in testability, modularity, commonality, etc.

In the **Advanced Development Phase**, the advanced design model is often a prototype (brassboard) of the system which not only works but also looks like the system. This is the state when the support system should be conceived and specified, at least at a basic level.

In the **Full-Scale Development Phase**, the engineers re-

PRODUCT LIFE CYCLE AND DEVELOPMENT LIFE CYCLE OVERLAP						
CONCEPT FORMULATION	CONTRACT DEFINITION	DEVELOPMENT		PRODUCTION	OPERATION	DISPOSAL
ENGINEERING DEVELOPMENT	ADVANCED DEVELOPMENT	FULL SCALE DEVELOPMENT	LOW RATE INITIAL PRODUCTION	FULL PRODUCTION		
ENGINEERING DEVELOPMENT MODEL	ADVANCED DEVELOPMENT MODEL	FORM/FIT/FUNCTION MODEL	PRODUCTION PROTOTYPE (PRODUCABLE)	PRODUCTION MODEL (CONTROLLED)		
"BREAD-BOARD"	"BRASS-BOARD"	TESTABILITY SUPPORTABILITY	ECO TO FIX FAULTS	ECO TO FIX FAULTS		
PROVE CONCEPT WORKS	MAKE IT FIT	BIT, BITE	. . .	. . .		
ESTABLISH SUPPORT CONCEPT & NEEDS	DEFINE SUPPORTABILITY	DEVELOP SUPPORT SYSTEM	PROVE SUPPORT SYSTEM	SUPPORT THE SYSTEM		
DEFINE EXTENT OF LSA/LSAR	ESTABLISH LSA/LSAR	IN-DEPTH LSA/LSAR	DETAILED LSA/LSAR	MAINTAIN LSAR		
ESTABLISH MANUAL CONCEPT	INITIATE MANUAL PLAN	WRITE DRAFT MANUAL	WRITE FINAL MANUAL	UPDATE MANUALS		

Figure 2-3 ILS in the Product Life Cycle

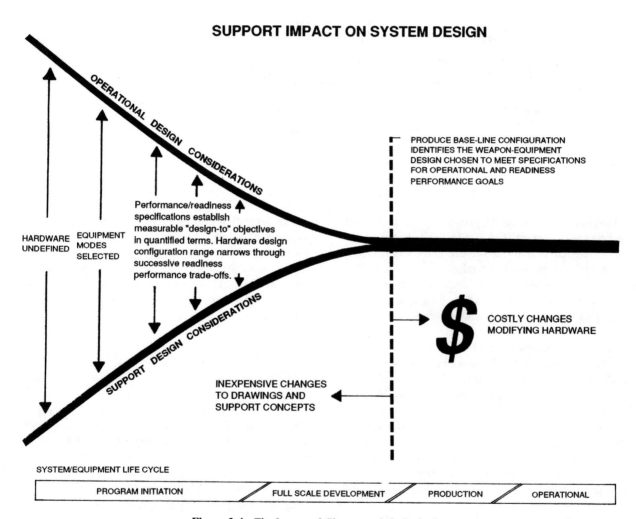

SUPPORT IMPACT ON SYSTEM DESIGN

OPERATIONAL DESIGN CONSIDERATIONS

PRODUCE BASE-LINE CONFIGURATION
IDENTIFIES THE WEAPON-EQUIPMENT
DESIGN CHOSEN TO MEET SPECIFICATIONS
FOR OPERATIONAL AND READINESS
PERFORMANCE GOALS

Performance/readiness
specifications establish
measurable "design-to" objectives
in quantified terms. Hardware design
configuration range narrows through
successive readiness
performance trade-offs.

HARDWARE
UNDEFINED

EQUIPMENT
MODES
SELECTED

SUPPORT DESIGN CONSIDERATIONS

$ COSTLY CHANGES
MODIFYING HARDWARE

INEXPENSIVE CHANGES
TO DRAWINGS AND
SUPPORT CONCEPTS

SYSTEM/EQUIPMENT LIFE CYCLE

| PROGRAM INITIATION | FULL SCALE DEVELOPMENT | PRODUCTION | OPERATIONAL |

Figure 2-4 *The Impact of Change on Life Cycle Cost*

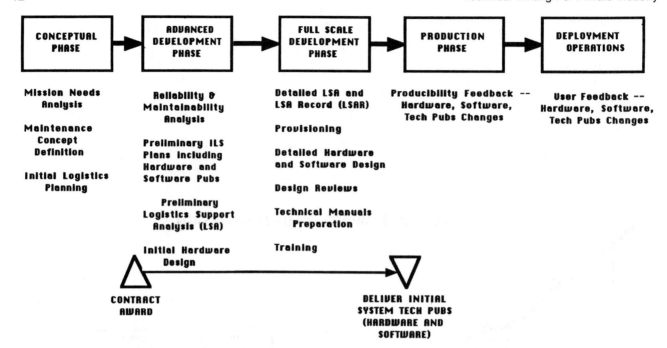

Figure 2-5 *Major Program Phases for Procurement*

vise the system so that it can be produced. This is the best time to enhance supportability and logistics design. After the support system design is developed in more detail, system supportability is demonstrated to the customer. The customer expects to have an early draft of the manual and documentation to use in evaluating the total system, so the manual often becomes a major and necessary part of the training program.

The extensive tests during the **Full Development Phase** and the needs of the **Full Production Phase** create the need for further changes. Since this is often the last cost-effective point to incorporate supportability modifications, changes during the **Low-Rate Initial Production Phase** may cause extensive revision of the manual that supports the system. You need to remember, though, that different branches of the services use slightly different approaches to acquire hardware. For instance, the operating guidelines of the Army would not totally describe the guidelines of the other services.

Figure 2-5 illustrates the five major program phases for procurement. Only the first two phases determine the need for technical publications. By the third phase, the publi-

cations are being generated, and by the end of the third phase they are available for use in the field. During the fourth and fifth phases, the field manuals are in use; any shortcomings are found and corrected at this time. In addition, hardware changes to modify equipment, for reasons such as modernizing or correcting defects, are also made in the manuals.

Conclusion

The technical writer's job does not end with production of the manual because through the economic life of the system there are changes—changes due to evolving use, changes due to design flaws, and changes because parts or subsystems become unsupportable in the field. In each case, the writer must be able to revise the manual. Only when the Disposal Officer accepts the system at the end of its life cycle does the writer's task end for a particular piece of hardware, software, or equipment. Today the emergence of concepts such as "pageless documents" and "paperless" manuals will change the writer's role in the development of the manual, but will not reduce the writer's importance.

Chapter 3

Military Specifications and Standards—How They Affect Technical Documents

Introduction

Why is regimentation necessary? Military Specifications (Mil Specs) and Standards (Stds) are guidelines the government has established to acquire material and data from contractors. The Mil Specs and Stds regiment technical publications and come in many variations, each designed to identify and satisfy unique requirements. Each branch of the service has developed its own specifications and standards to meet its own needs, but at times each branch can use the same specifications and standards, thus reducing the need for multiple versions. This chapter describes the contents of a format/style specification *"Manuals, Technical: General Style and Format Requirements"* (MIL-M-38784) and a content specification *"Manuals, Technical: Commercial Equipment"* (MIL-M-7298), as examples of typical Military Specifications.

The Need for Regimentation

The need for regimentation or sameness is necessary in technical publications for any given type of manual (Operator, O&M, Depot Maintenance Work Request (DMWR), etc.). Manual users should be able to find similar information in the same place, formatted in the same manner, and written in the same style. Instructors need manuals that are uniform to train soldiers to use equipment, and soldiers need to reference numerous technical manuals for information on various hardware items. For example, Paragraph 3.3 of MIL-M-38784 states the requirements for uniformity:

For maximum clarity and usefulness, there shall be consistency in terminology within the same publication or series of publications. To the extent that the nature of the data being presented will allow, there shall be consistency of organization among like-type publications.

Types of Specifications and Standards

Today specifications and standards number in the thousands. Many pertain to technical publications, but of those, only a few are cited over and over again in contracts. There are many categories of specifications and standards, including the following:

- Calibration
- Checklists
- Commercial
- Corrosion control
- Damage control
- Data
- Definitions
- Drawing
- Inspections
- Installation
- Lubrication
- Maintenance
- Management
- Marketing measurement
- Nomenclature
- Operation
- Ordinance
- Overhaul
- Parts lists
- Production and format
- Provisioning
- Quality
- Repair
- Rescue
- Safety
- Schedules

Technical publication specifications are divided into two types:

- Format/Style Specifications
- Content Specifications

Any document you read will contain certain aspects of those two types. But of all the specifications and standards, as a technical writer you will normally use only a few; it all depends on the company you work for, the type of product, and the customers' needs. For each type of manual, most contracts require only one or two writing specifications (content and format), some standards for illustrations, symbols, and abbreviations and a printing specification.

For example, the Applicable Documents paragraph of both MIL-M-38784 and MIL-M-7298 includes lists of specifications and standards which are typical of those you must be familiar with. Remember, each branch of the government has its own specs, as well as some that are common to all branches. Table 3-1 gives only a very brief listing of some of the more commonly used specs and standards (both military and commercial).

Specifications are periodically updated to keep up with user needs. With each update, a new change identification shows the latest release. Examples of MIL-M-38784 and MIL-M-7298 contained in this book are taken from the "B" version of MIL-M-38784, and the "C" version of MIL-M-7298.

Format/Style Specifications

Format and style specifications dictate the way material will be presented in a manual, such as how to break up a document into chapters, sections, paragraphs, and subparagraphs. They dictate how material is to be prepared and presented, including the following:

- Numbering systems for chapters, sections, pages, paragraphs, tables, and figures
- Format (set up) of tables
- Image area of a page (borders, number of columns) and page size
- Printing requirements
- Typing requirements (font, point size, justification, non-justification)
- Front matter requirements (cover, title page, table of contents, list of illustrations, foreword)
- Requirements for warnings, cautions, and notes

You need to remember that specification requirements can be superseded by deviations cited in the contract (usually within the Statement of Work, or SOW).

MIL-M-38784, Manuals, Technical: General Style and Format Requirements

MIL-M-38784, ''Manuals, Technical: General Style and Format Requirements,'' is a typical format specification used in preparing data for the government. It is cited in most U.S. Army, Navy, Air Force, and Marine contracts where requirements exist for technical manuals.

MIL-M-38784 is broken down into six key paragraphs and contains numerous examples of format requirements:

Paragraph 1 - Scope
Paragraph 2 - Applicable Documents
Paragraph 3 - Requirements
Paragraph 4 - Quality Assurance Provisions
Paragraph 5 - Preparation for Delivery
Paragraph 6 - Notes

Figures 1 through 40 at the end of this spec (MIL-M-38784) give examples of the required format and style. An index is provided for the specification. Blank DD Form 1426, Standardization Document Improvement Proposal, is provided for those agencies initiating technical data procurements which have identified deficiencies in the spec and recommend changes (another addendum to this spec).

Analysis of MIL-M-38784

MIL-M-38784 is designed to meet the requirements of all the military services needing a basic format spec. However, as already mentioned, there are exceptions where one or more of the services may have unique requirements. In such cases, parenthetical letters (A) for Army, (N) for Navy, (MC) for Marine Corps, and (F) for Air Force are used. When a paragraph is preceded by one or more of these parenthetical references, that paragraph applies only to those services noted.

Paragraph 3. Requirements

Paragraph 3. of MIL-M-38784 is divided into the following ten main areas:

- *Paragraph 3.1.1 Conflict Between Documents*: Due to variations in the requirements, quite often there are conflicts between the numerous documents cited in the contracts. These conflicts may be either major or minor. Paragraph 3.1.1 of this spec and Table 3-2 in this chapter define the order of precedence the requirement or requirements must follow (you should be aware that this order of precedence may be superseded by contract stipulations).

- *Paragraph 3.1.4 - Manual Outline*: When specified by the contract, an outline should be prepared In Accordance With (IAW) the requirements of Paragraph 3.1.4. The outline has to identify paragraphs, figures, and tables expected in the manual. In addition, the estimated page count of each chapter has to be provided. Once the outline is reviewed and accepted by the customer, it becomes the guiding document, the master plan for preparing the manual. Technical writers are not free to deviate from this plan unless modifications to an outline are approved by the customer. As a first step, all deviations should be evaluated by the writer's supervisor and possibly the program manager prior to seeking customer approval.

- *Paragraph 3.2 - Format*: The requirements for both manuscript and camera-ready copy are defined in Paragraph 3.2. Specific guidelines are given for page size, typing requirements, image area available for different sizes of manuals, and requirements for oversized copy. Guidelines also cover numbering of pages, paragraphs, tables, and illustrations. The handling of foldouts, charts, tables, footnotes, and illustrations, as well as subordination of paragraphs and procedural data, and creating appendices and glossaries are described.

- *Paragraph 3.3 - Style of Writing*: General guidelines for writing are given in Paragraph 3.3, including references, person and mood, readability, and use of abbreviations and acronyms. The differences in a warning, a caution, and a note are also defined.

- *Paragraph 3.4 - Security, Classification Requirements*: Requirements for proper identification of classified material are given in Paragraph 3.4 and are very stringent. Since you are dealing with the national defense, the Department of Defense (DOD) is extremely cautious in preparing and dispersing classified material.

- *Paragraph 3.5 - Front Matter*: How to prepare front matter (those pages from the cover to the first chapter) is extensively covered in Paragraph 3.5. Instructions to prepare the cover and title pages alone take up five pages, including three pages of illustrations. Requirements for uniformity are stressed.

- *Paragraph 3.6 - Illustrations*: The requirements for preparing illustrations (photographs, line drawings, and

Table 3-1. *Typical Military/Commercial Specifications and Standards*

SPEC/STD	Title
MIL-STD-12	Abbreviations for use on drawings, Specifications and Standards in Technical Documents
USAS Y14.15	Electrical and Electronics Diagrams
ANSI Y32.2	Graphic Symbols for Electrical and Electronics Diagrams
ANSI Y32.14	Graphic Symbols for Logic Diagrams (Two-State Devices)
ANSI Y32.16	Reference Designations for Electrical and Electronics Parts and Equipment
DOD-STD-100	Engineering Drawing Practices
ANSI/IEEE STD 260	IEEE Standard Letter Symbols for Units of Measure
MIL-HDBK-63038-1	Technical Manual Writing Handbook
MIL-HDBK-63038-2	Technical Writing Style Guide
MIL-M-7298	Manuals, Technical: Commercial Equipment
MIL-M-63041	Manuals, Technical: Content Requirements for Depot Maintenance Work Requirements
MIL-M-81260	Manuals, Technical: Aircraft Maintenance
MIL-M-82376	Manuals, Technical: for Training Devices
DOD-D-1000	Drawings, Engineering and Associated Lists
MIL-STD-4752	Reading Level Requirements for Preparation of Technical Orders
MIL-M-15071	Manuals, Technical: Equipments and Systems Content for
MIL-M-21548	Manuals, Technical, FBM Weapon System, General Specification for
MIL-M-24100	Functionally Oriented Maintenance Manuals (FOMM) for Equipment and Systems
MIL-M-38717	Manuals, Technical: Work Unit Code (for Ground Communications Electronics, Meteorological Equipment)
MIL-M-38777	Manuals, Technical: Inspection Requirements, Lubrication Requirements and Word Cards for Ground Communication, Electronic and Meteorological (CEM) and Related Equipment
MIL-M-38778	Checklist, Title Page, List of Effective Pages, Printing and Binders, General: Requirements for Preparation of
MIL-M-38784	Manuals, Technical: General Style and Format Requirements
MIL-P-38790	Printing Production of Technical Manuals: General Requirements of
MIL-M-38793	Manuals, Technical Calibration Procedures, Preparation of
MIL-M-38798	Manuals, Technical: Operation Instructions, Maintenance Instructions, Circuit Diagrams, Alignment Procedures and Installation Planning
MIL-T-38804	Time Compliance Technical Orders (TCTO's) Preparation of
MIL-M-388807	Manuals, Technical: Illustrated Parts Breakdown, Preparation of
MIL-M-63000	Manuals, Technical: General Requirements for Manuscripts
MIL-M-63001	Manuals, Technical: Repair, Parts and Special Tools List
MIL-M-63002	Manuals, Technical: Requirements for Modification Work Orders
MIL-M-63004	Manuals, Technical: Preparation of Lubrication Orders
MIL-M-63036	Manuals, Technical: Operator's Preparation of
MIL-M-63038	Manuals, Technical: Organizational or Aviation Unit, Direct Support or Aviation Intermediate, and General Support Maintenance.

Table 3-2 *Order of Precedence*

Conflict Between	Precedent Document
Contract and MIL-M-38784	Contract
Contract and governing content spec	Contract
MIL-M-38784 and content spec	Governing content spec
MIL-M-38784 and its referenced specs	MIL-M-38784
U.S. Government Printing Office Style Manual and MIL-M-38784 or any other MIL Standard format requirement	Format requirement of MIL-M-38784 or any other MIL Standard format requirement

combinations of both) are given in Paragraph 3.6. This paragraph covers just about every area of art, including the drawing, mounting, identifying, and covering (flapping) of art.

● *Paragraph 3.7 - Changes*: Changes to both manuscripts and camera-ready copy (CRC), an important aspect of technical publications, are covered in Paragraph 3.7. Manuals are, in effect, living documents because they are always kept current. When a piece of hardware becomes outdated or obsolete and modifications are incorporated into the current hardware to update or correct design errors, the technical publications must also be updated to reflect these changes. When a change is made and the changed pages are distributed to all the holders of the document, the old pages must be removed and the new ones inserted. The size of the impact on the existing document determines whether a "change" or a "revision" will be necessary. Paragraph 3.7 defines a change as follows:

A change is comprised of corrected pages to the basic manual. It consists of information that improves or clarifies the basic manual without requiring rewriting or reorganization of the technical contents of the basic manual.

The alternative to a change is a revision, which is covered in Paragraph 3.8.

● *Paragraph 3.8 - Revisions*: When extensive changes are required such that the majority of pages are effected, it is senseless to have this large a number of pages identified as change pages. According to Paragraph 3.8, a revision " . . . shall be prepared when specified in the contract or order." There is no specific breaking point when a revision rather than a change would be required. The general guideline is to estimate the percentage of new or revised material over the original material.

Generally, a modification affecting between 50 and 60 percent or higher of the original document will require a revision. Paragraph 6.3.23 of the same spec states, "A revision is a second or subsequent edition of a manual which supersedes the preceding edition." It then differentiates between an "update revision" and a "complete revision." Both require that the entire manual be reissued.

● *Paragraph 3.9 - Supplements*: In some cases, a third alternative to preparing new material for an existing document is used. This alternative is called a "supplement" and is covered in Paragraph 3.9. According to Paragraph 6.3.26 of the same spec, "A supplement is a subsidiary document which complements information in a manual." The supplement provides additional data which is determined to be better suited to a separate document rather than being included in an existing document, as for example, the classified supplement. The supplement to the manual, properly classified, can be made available only to those who need to know that information. Those who need to be restricted from accessing the entire document can be denied access to the supplement. Limiting the availability of classified information in this manner decreases the chances of having that information fall into the "wrong" hands.

● *Paragraph 3.10 - Brief Manuals*: A manual having eight or less pages is called a brief manual and is described in Paragraph 3.10. Because of its small size, there are some minor restrictions to the front matter, formatting, and use of revisions rather than changes.

Paragraph 4. - Quality Assurance: Paragraph 4. defines the requirements for inspection, quality conformance, government inspection, and readability. Customers want to make sure they are getting a document which meets their requirements. Paragraph 4. discusses the requirements for inspecting the manual to determine how well it was prepared according to the specification. It also defines the method to determine the manual's readability. Readability is a key issue because all manuals are prepared to meet the specific training levels of the users.

Paragraph 5. - Preparation for Delivery: Paragraph 5. defines the requirements that contractors must use for delivering the manuscript and the camera-ready copy, and preparing classified material and original artwork. It also discusses how to package the material and how to use packing lists to check the materials being delivered.

Packaging classified data is given only minor treatment and references DOD 5220.22-M as a guideline. When handling classified information, specific requirements must be adhered to. Technical writers must read all pertinent specifications and become familiar with not only all the shipping requirements but also the day-to-day handling and preparing of such material. Writers should always be fully aware of security requirements stipulated by their employer's security organization.

Content Specifications

Content specifications define how chapters and sections should be arranged, including requirements for physical, operational, and functional descriptions. They cite specific requirements for types of tables, sometimes even identifying column headers. For example, MIL-M-7298, "Manuals, Technical: Commercial Equipment," is a typical technical publications content spec and is designed to define the government's minimum requirements for an acceptable commercial manual.

In general, the requirements for a commercial manual are not as stringent as the requirements for a military manual. The commercial manual is a versatile document, designed to fit the needs of various technical documentation requirements. The commercial manual spec is very general. It only describes what should be contained in the chapters and sections (i.e. Introduction, Installation, Operation, etc.). Since it is generic it provides guidelines on what should be said about the hardware; it does not mention any specific hardware.

MIL-M-7298: Content Specification

Paragraph 1.1.1 of MIL-M-7298 defines "the minimum requirements for acceptable commercial manuals for use by all departments and agencies of the government to install, operate, and maintain commercial equipment."

The government purchases two types of hardware: those specifically designed and built for the government using government funds, and those which are built commercially and sold over-the-counter or "off-the-shelf" (OTS), also referred to as "Commercial off-the-shelf," or (COTS). When a contractor is building a piece of hardware for the government, the technical manual written to describe the equipment must meet the requirements of the Military Specifications cited in the contract. However, when a company manufactures a company funded OTS item, a computer for instance, that company does not have to write a technical manual to anyone's standards except its own.

The technical manual format and content standards defined by the company may result in a technical manual that does not meet the minimum requirements of the government. If the government (or prime contractor) compares an OTS manual against the requirements of MIL-M-7298, and determines the need to supplement that manual, then MIL-M-7298 will be a guideline for such supplements.

The spec is arranged similar to MIL-M-38784. Its key paragraphs are as follows:

- Scope
- Applicable Documents
- Requirements
- Quality Assurance Provisions
- Preparation for Delivery
- Notes

The key part of the spec is Paragraph 3 (Requirements), which is broken down into four main areas:

- Acceptable Commercial Manual
- Supplemental Data
- Copyrights
- New Manuals

Similar to MIL-M-38784, the paragraphs of MIL-M-7298 are keyed to military branches: (A), (N), (MC), and (F) or (ALL), meaning for all branches.

Paragraph 3.1 - Acceptable Commercial Manual: The acceptable commercial manual may be simple or complex. The simplicity of the equipment determines the content of the manual, but a complex manual contains more extensive coverage, including the following:

- Front Matter
- Table of Contents and List of Illustrations
- Safety Precautions
- Introduction
- Preparation for Use
- Principles of Operation
- Operation Instructions
- Maintenance and Servicing Instructions (Preventive and Corrective)
- Overhaul Instructions (as applicable)
- Preparation for Shipment
- Storage
- Parts List
- Illustrations and Diagrams

Each of these subjects may be an entire chapter or some subjects can be separated into several chapters entitled, Cleaning and Lubrication, Troubleshooting, Performance Verification, and Maintenance Instructions. Generally, Safety Precautions is not a chapter but is (1) summarized in the beginning of the book and (2) stated wherever applicable as cautions and warnings.

Differences in the basic names of the chapters from one spec to another do not mean that the contents of the chapters are not similar. The Introduction Chapter in one spec may contain the same type of information as the Description Chapter in another. Similarly, Preparation of Use and Installation Instructions in MIL-M-7298 may have the same content as Installation in another spec. The chapter titles are arbitrary.

Paragraph 3.2 - Supplemental Data: When the basic content of a commercial manual is satisfactory but is lacking in certain areas, a supplement to the manual may be needed. This paragraph gives information on how to prepare supplementary information.

Paragraph 3.3 - Copyrights: This paragraph provides information on the use of copyright data and the need for copyright releases.

Paragraph 3.4 - New Manuals: A new manual may have to be written if no commercial manual exists or if supplemental data to an existing manual would be inadequate. This paragraph provides information on how to prepare a new manual to either MIL-M-7298 or some other specification cited in the contract.

Technicians must be given certain key types of CONTENT information. They must know what the hardware is, what it looks like, what physical and functional characteristics it has, how to install and remove it, how to ship it, how to operate it, how it works (theory), and how to maintain it (including testing, disassembling, adjusting, lubricating, calibrating, repairing, and reassembling). They also need pictures and diagrams showing where everything is located, what the parts are, and how it is put together. Therefore, the breakdown of information in this spec is basically the breakdown of almost every technical manual content specification.

Conclusion

Military specifications provide technical writers with information on what to write and how to write it. The large variety of Mil Specs can be divided into two basic types:

format and content. MIL-M-7298 is a government specification which identifies the requirements for acceptable commercial manuals. One format spec, MIL-M-38784, has been established as an accepted authority which is used for most technical documents procured by the government. The government also distinguishes between formal Mil Spec formatted manuals and commercial manuals. All technical writers using these specs must understand how to interpret them. But there is a great similarity in the format of the spec requirements. Once familiar with their overall structure, the writer's task of preparing text and illustrations based on accurate interpretation is simplified.

Chapter 4

The Procurement Process

Introduction

Once the government (the customer or buyer) recognizes a need for acquiring a piece of equipment (hardware), it initiates a solicitation which may be in the form of a Request for a Proposal (RFP), a Request for Quote (RFQ) or an Invitation for Bid (IFB). The solicitation is directed to industry or companies (the contractors), which in turn respond with a formal proposal which may lead to a contract.

The RFP, RFQ, or IFB defines the design concept of the hardware wanted, identifies the required specifications and the quantity of deliverable hardware items, and gives specific dates for scheduled milestones and deliveries. The solicitation also contains a Statement of Work (SOW) which further describes the planned acquisitions and a series of Contract Data Requirements Lists (CDRL) items that specify particular guiding documents and identify the quantities of data to be delivered, the scheduled due dates, and the agencies to which the deliverable data is to be sent.

Procurement Types

Generally, the difference in an RFP, RFQ, and IFB is the amount of risk to the bidders or contractors. The IFB is used when there are few technical characteristics that are not clearly defined. Detailed information has been developed (the specifications), and the contractors face only slight risks to develop the new "system." IFB procurements have a public opening of the sealed bids, and the results are usually known immediately. In addition, the contracts for IFB procurements are Firm Fixed Price (FFP). The RFQ is similar to the IFB in that a degree of uncertainty exists for those contractors who wish to bid.

On the other hand, the RFP is used when a degree of technical uncertainty does exist; the contractor may knowingly take on some factor of risk. Negotiations are either competitive, with the best and final offer winning, or sole source (bid offered to just one contractor) along with the required fact finding and negotiation process.

Technical Input to the Proposal

The contractor responds to the RFP, RFQ, or IFB with a proposal (the contractor's Technical Publication) which covers the technical, management, and cost information. The technical part of a proposal describes how the contractor plans to meet the obligations defined by the solicitation and may describe the company's history of similar contracts awarded to the contractor. It identifies key personnel who will be assigned to do the work in the time required, and the capital equipment and facilities that will be available to do the job. It describes how the contractor will manage the job, will monitor the cost and schedule, and identifies how the contractor interprets all the requirements.

Contractor's Organization

In a proposal, information on a company's organizational structure should include a description of the functions and responsibilities of the Technical Publications' task leader. It should describe how the task leader will control the preparation, quality, delivery schedule, and cost of the technical manual.

Previous Experience

Contractors must prove to the customer that they are competent to perform the tasks required in the contract. To do this, contractors include their work history or "track record" in the proposal. It is a definite plus if the contractors can list previous work done for various government agencies, including many different types of technical documents for numerous types of equipment. Also having adequate and experienced personnel can make the difference between showing the customer their company's ability to perform versus questionable or marginal ability to perform. Resumes of key personnel to be assigned to the program may also be included in the proposal.

Task Performance

A major element of the proposal is convincing the customer that the contractor understands and is able to perform the task. Contractors must persuade the customer that their personnel have the experience and talent to get the job done properly, on schedule, and within the prescribed budget.

Program Milestones

The proposal's schedule identifies major program milestones. Figure 4-1 illustrates a typical example of a simple Technical Publications' milestone schedule for several typical tasks. The tasks identified vary from one contract to another and depend upon requirements cited in the CDRL's and SOW. Delivery requirements can be stated by various methods. For example, After Date Award Document (ADAD)

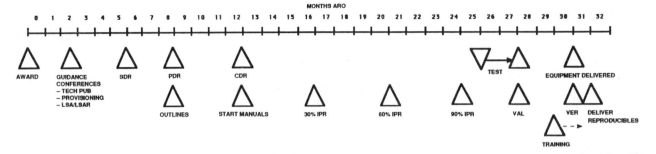

Figure 4-1 *Schedule of Technical Publications Milestones*

or contract award date represents day 1 of the contract. Manual related requirements such as the CDRL may state that the manuals are due a certain number of days ADAD (e.g., 60), which means they are due to the customer that number of days after the contract is awarded. Contract dates may be identified relative to Months After Contract (MAC), Days After Contract (DAC), After Receipt of Order (ARO), or Equipment Delivery (ED) to the customer.

As shown in Figure 4-1, Months ARO are listed horizontally. In this example the technical publications program lasts 31 months, but the deliverable hardware part of the program may stretch out several years for a large production contract. The program milestones are shown as the top row of triangles, and the technical publications task is shown as the second row of triangles.

Cost Input to the Proposal

The technical publications cost portion of the proposal used to be provided to the customer on DD Form 633-2, DOD Contract Pricing Proposal (Technical Publications) based on the Defense Acquisition Regulations (DAR). This form (see Figure 4-2) provided the customer with information about technical data cost elements. Page 1 provided fill-in spaces for the contractor's summary estimates on work hours, total cost, cost related reference material (basis of estimates), and unit counts on pages of text and illustrations. Page 2 contained instructions to contractors (offerors) about filling out the forms and a list of questions related to their ability to perform with their own funds.

A new Contract Pricing Proposal Cover Sheet, Form 1411 (Figure 4-3), has replaced the DD Form 633-2 and is based on the new Federal Acquisition Regulations (FAR). The new form does not require and does not provide space for page unit details.

Estimating Costs

The cost of technical manuals is based on estimating the sum cost of three factors: (1) number of pages (text/tabular, and illustrations); (2) cost of labor per hour (technical writer, illustrator, word processor, editor, engineer, etc.) to write and illustrate the pages, and (3) cost of materials (printing, negatives, stats, covers, binders, etc.). To this basic cost is

added the company's overhead (O/H) and general and administrative (G&A) costs.

Estimating Page Count

All inputs to the proposal for costing technical publications are based on the manual's size, which entails the number of pages of text, tables and illustrations in the manual. Page counts for text are typically typed single space, on an 8 1/2 × 11 inch page. Both text (prose) and tabular material are considered to be text in a proposal. Illustrations may be of various types and sizes. Estimating the page count in the manual can be done by the automated tabulation, the specification breakdown or the hardware method. Each is described in more detail in the following paragraphs.

• Estimating Page Count by Automated Tabulation

Some companies automate their bidding process by entering certain basic parameters such as page counts, costs, and backup data into an algorithm to come up with the "basis of estimate" (BOE). These basic parameters then are automatically tabulated. A BOE provides the customer with statistical data on how estimates were derived; it justifies the hour and dollar calculations. Advantages to using the Automated Tabulation method are the consistency obtained between bids and ease of making changes.

• Estimating Page Count by Specification Breakdown

The Specification Breakdown method of estimating page count typically uses an outline of the technical manual on the chapter and section level. More detailed outlines can be used if the bidder finds it necessary to be more precise. Figure 4-4 is a typical example of a form that can be used to estimate page counts and identify other characteristics of the bid. The page columns provide space to account for text and tabular data, halftone photographs, and full page unit (8 ½ × 11 inch) line drawing illustrations, and foldout (greater than 8 ½ × 11 inch and typically 11 × 17 inch page size) functional and engineering illustrations. The example uses MIL-M-38798 as the guiding specification. The bidder's outline will be derived from specifications cited in the solicitation's SOW and/or CDRL DD Form 1423. After preparing the outline to the desired level, the bidder then equates each of the outline's parts with the number of pages needed to describe the hardware. The bidder's knowledge of the hardware is crucial to arrive at an accurate estimate of the page count.

● Estimating Page Count by Hardware Breakdown

The Hardware Breakdown method of estimating page count is based primarily on hardware type and complexity. In addition to having a firm understanding of the applicable Mil Specs, the bidder must have experience interpolating page counts based on:

1. Complexity of controls and indicators
2. Total number of and types of repairable/throw-away assemblies referred to as Line or Lowest Repairable Assemblies (LRU's)
3. Circuit card density (number and type of components on each card type)
4. Maintenance concept (customer repairable, vendor repairable, or throw-away assemblies and subassemblies)
5. Extent of built-in test/built-in test equipment (BIT/BITE)
6. Test equipment requirements
7. Make-buy hardware (commercial off-the-shelf items to be used "as-is" or to be modified, or items to be built by the contractor)

The type of manual to be written also determines how extensive the information should be. If you are bidding on an operator's manual, the primary concern for page count estimating will be the complexity of controls and indicators, operator interface (i.e. computer driven displays) and operator level maintenance. An Operation and Maintenance (O&M) manual requires much greater detail and more extensive information due to the need for data on theory (how the equipment works) and more complex maintenance activities. This bidding method relies upon the bidder's ability to equate page counts to types of hardware. The estimate is based on either of two approaches.

In the first approach, the bidder makes an assumption about the number of pages for front matter and "boilerplate" material. Physical description pages (a description of what the unit is, what it contains, where the parts are located, and with what parameters it operates) must also be accounted for in text, tables, and illustrations. Identification of operator controls and indicators and their use (operating information) require additional text, tabular, and illustrative pages based on quantity and complexity of controls and indicators. An estimate must be made of text, tabular, and illustrative material required to describe operation of circuit cards based on whether the cards have light, medium, or high density of components mounted thereon. Similarly, text and illustration pages for preventive and corrective maintenance must be accounted for based on hardware design factors such as amount of built-in test equipment, number of adjustable components, etc.

Supporting information in the form of engineering schematic, wiring, cabling, etc., must also be included. Wire run lists may be used in place of wiring/cabling diagrams if they will meet contract requirements. If the circuit cards are repairable, schematic diagrams usually must be included for each unique type of card. (Data in the technical manual should always agree with the maintenance concept). The quantity of sheets for circuit card schematic or logic diagrams (usually "B" size or 11 × 17 inch size engineering drawings) will vary depending upon card density and could require five or more pages per card. Finally, if you are including special (depot) level maintenance data, the quantity of pages will depend upon whether card tests are done manually or done by using some form of Automatic Test Equipment (ATE).

In the second approach, the hardware breakdown method, a rougher estimate than the one just described, may be used which is based strictly on the number of controls and indicators, assemblies (LRU's) and types of assemblies in the unit, their complexity, and the maintenance concept. All of the various page types required throughout the manual are rolled into one average number of pages per circuit card (LRU), which is then multiplied by the number of circuit card types. Thus, the bidder may estimate 10, 20, 30, or 40 pages per LRU, multiply that number by the total number of LRU's (or LRU types) to come up with a very rough page estimate. When using any of these estimating methods, you can still expect differences in the total page count.

Page Counts: An Inexact Process

Page count estimating is a very inexact process. There are several major reasons for this. First and foremost is the bidder's experience. A person having extensive experience writing manuals to a large variety of specifications for numerous customers will have a better understanding of the writing process than one who lacks that experience. The knowledgeable person will therefore be better able to understand and interpret contract requirements. A second reason is the bidder's understanding of the hardware about which the manual will be written. The number of pages needed to describe equipment in the detail cited by applicable specifications is dependent upon the size, complexity, operability, and maintainability of that equipment.

A third reason for the inexact page count estimate is knowledge of the customer's requirements. A particular piece of hardware might require a 200-page technical manual to satisfy the needs of one customer and a 100-page manual for another customer. The technical publications page counts of competitive bids submitted by various companies for any given piece of equipment might vary by as much as 100 percent. If you have previously prepared technical manuals for a particular customer, you possess an insight into how the customer interprets specifications, and what the customer looks for and expects from contractors in both quantity and quality of technical documentation. If this potential customer is new to your company but is known by your competition, you are at a disadvantage.

In the hardware breakdown method you will be losing sight of the specifics of what text, tables, and illustrations

DEPARTMENT OF DEFENSE **CONTRACT PRICING PROPOSAL** *(Technical Publications)*	Form Approved Budget Bureau No. 22-R100	
This form is for use when *(i)* submission of cost or pricing data *(see ASPR 3-807.3)* is required and *(ii)* substitution for DD Form 633 is authorized by the contracting officer.	PAGE NO.	NO. OF PAGES

NAME OF OFFEROR	PUBLICATION TITLE
HOME OFFICE ADDRESS *(Include ZIP Code)*	SPECIFICATION NO. OR OTHER DESCRIPTION
DIVISION(S) AND LOCATION(S) WHERE WORK IS TO BE PERFORMED	GOVT SOLICITATION NO. / SOLICITATION LINE ITEM NO.

SUMMARY

A. COST ELEMENTS	MAN HOURS	TOTAL COST[1]	REFERENCE[2]
1. DIRECT MATERIAL[3]			
a. PURCHASES[4]			
b. SUBCONTRACTS *(excluding printing & binding)*[5]			
c. SUBCONTRACTS *(printing and binding)*[6]			
2. MATERIAL OVERHEAD[7]			
3. DIRECT LABOR[8]			
a. TEXT[9]			
b. ILLUSTRATING[10]			
c. COPY PREPARATION[11]			
d. OTHER[12]			
4. LABOR OVERHEAD[7]			
5. OTHER COSTS[13]			
6. *SUBTOTALS*			
7. GENERAL AND ADMINISTRATIVE EXPENSE[7]			
8. *SUBTOTALS*			
9. PROFIT OR FEE			
10. **TOTAL PRICE** *(AMOUNT)*			

B. ESTIMATED PAGE UNITS[14] *(per publication)*	NEW		REVISED		UNCHANGED		TOTALS	
	UNITS	MAN HOURS	UNITS	MAN HOURS	UNITS	MAN HOURS	UNITS	MAN HOURS
1. TEXT:								
a. NARRATIVE								
b. PARTS LISTING								
2. ILLUSTRATIONS:								
a. TO NARRATIVE								
b. TO PARTS LISTING								

This proposal is submitted for use in connection with and in response to _____

_____ * and reflects our best estimates as of this date, in accordance with the instructions to offerors and the footnotes which follow.

    DESCRIBE RFP, ETC.

TYPED NAME AND TITLE	SIGNATURE
NAME OF FIRM	DATE OF SUBMISSION

DD FORM 1 APR 68 **633-2**   PREVIOUS EDITIONS ARE OBSOLETE.

Figure 4-2 *DD Form 633-2, DOD Contract Pricing Proposal*

INSTRUCTIONS TO OFFERORS

1. The purpose of this form is to provide a standard format by which the offeror submits to the Government a summary of incurred and estimated cost *(and attached supporting information)* suitable for detailed review and analysis. Prior to the award of a contract resulting from this proposal the offeror shall, under the conditions stated in ASPR 3-807.3, be required to submit a Certificate of Current Cost or Pricing Data *(see ASPR 3-807.3(e) and 3-807.4).*

2. As part of the specific information required by this form, the offeror must submit with this form, and clearly identify as such, cost or pricing data *(that is, data which is verifiable and factual and otherwise as defined in ASPR 3-807.3(e)).* In addition, he must submit with this form any information reasonably required to explain the offeror's estimating process, including:

    a. The judgmental factors applied and the mathematical or other methods used in the estimate including those used in projecting from known data, and

    b. The contingencies used by the offeror in his proposed price.

3. When attachment of supporting cost or pricing data to this form is impracticable, the data will be specifically identified and described *(with schedules as ap-*

propriate), and made available to the contracting officer or his representative upon request.

4. The formats for the "Cost Elements" and the elements of "Man Hours", "Total Cost" and "Reference" are not intended as rigid requirements. These may be presented in different format with the prior approval of the contracting officer if required for more effective and efficient presentation. In all other respects this form will be completed and submitted without change.

5. By submission of this proposals offeror, if selected for negotiation, grants to the contracting officer, or his authorized representative, the right to examine, for the purpose of verifying the cost or pricing data submitted, those books, records, documents and other supporting data which will permit adequate evaluation of such cost or pricing data, along with the computation and projections used therein. This right may be exercised in connection with any negotiations prior to contract award.

6. A separate DD Form 633-2 shall be submitted for each publication unless prior approval for consolidated submission is obtained from the contracting officer.

SECTION A - QUESTIONS

I. HAVE THE DEPARTMENT OF DEFENSE, NATIONAL AERONAUTICS AND SPACE ADMINISTRATION, OR THE ATOMIC ENERGY COMMISSION PERFORMED ANY REVIEW OF YOUR ACCOUNTS OR RECORDS IN CONNECTION WITH ANY OTHER GOVERNMENT PRIME CONTRACT OR SUBCONTRACT WITHIN THE PAST TWELVE MONTHS?

☐ YES ☐ NO   *IF YES, IDENTIFY BELOW.*

NAME AND ADDRESS OF REVIEWING OFFICE (Include ZIP Code)	TELEPHONE NUMBER

II. WILL YOU REQUIRE THE USE OF ANY GOVERNMENT PROPERTY IN THE PERFORMANCE OF THIS PROPOSED CONTRACT?

☐ YES ☐ NO   *IF YES, IDENTIFY ON A SEPARATE PAGE.*

III. DO YOU REQUIRE GOVERNMENT CONTRACT FINANCING TO PERFORM THE PROPOSED CONTRACT?

☐ YES ☐ NO   *IF YES, IDENTIFY:*   ☐ ADVANCE PAYMENTS   ☐ PROGRESS PAYMENTS   OR ☐ GUARANTEED LOANS

IV. WOULD ANY OF THE DIRECT COSTS SHOWN ON THIS FORM BE INCURRED UNDER THIS PROPOSED CONTRACT IF THESE TECHNICAL PUBLICATIONS WERE NOT A REQUIREMENT OF THE ABOVE CITED GOVERNMENT SOLICITATION?

☐ YES ☐ NO   *IF YES EXPLAIN ON A SEPARATE PAGE.*

V. DOES THIS COST SUMMARY CONFORM WITH THE COST PRINCIPLES SET FORTH IN ASPR, SECTION XV *(see 3-807.2(c)(2))?*

☐ YES ☐ NO   *IF NO, EXPLAIN ON A SEPARATE PAGE.*

Figure 4-2 *Continued*

CONTRACT PRICING PROPOSAL COVER SHEET	1. SOLICITATION/CONTRACT/MODIFICATION NO.	FORM APPROVED OMB NO. 3090-0116

NOTE: This form is used in contract actions if submission of cost or pricing data is required. *(See FAR 15.804-6(b))*

2. NAME AND ADDRESS OF OFFEROR *(Include ZIP Code)*	3A. NAME AND TITLE OF OFFEROR'S POINT OF CONTACT	3B. TELEPHONE NO.

4. TYPE OF CONTRACT ACTION *(Check)*	
A. NEW CONTRACT	D. LETTER CONTRACT
B. CHANGE ORDER	E. UNPRICED ORDER
C. PRICE REVISION/ REDETERMINATION	F. OTHER *(Specify)*

5. TYPE OF CONTRACT *(Check)*

☐ FFP ☐ CPFF ☐ CPIF ☐ CPAF

☐ FPI ☐ OTHER *(Specify)*

6. PROPOSED COST *(A+B=C)*		
A. COST	B. PROFIT/FEE	C. TOTAL
$	$	$

7. PLACE(S) AND PERIOD(S) OF PERFORMANCE

8. List and reference the identification, quantity and total price proposed for each contract line item. A line item cost breakdown supporting this recap is required unless otherwise specified by the Contracting Officer. *(Continue on reverse, and then on plain paper, if necessary. Use same headings.)*

A. LINE ITEM NO.	B. IDENTIFICATION	C. QUANTITY	D. TOTAL PRICE	E. REF.

9. PROVIDE NAME, ADDRESS, AND TELEPHONE NUMBER FOR THE FOLLOWING *(If available)*

A. CONTRACT ADMINISTRATION OFFICE	B. AUDIT OFFICE

10. WILL YOU REQUIRE THE USE OF ANY GOVERNMENT PROPERTY IN THE PERFORMANCE OF THIS WORK? *(If "Yes," identify)* ☐ YES ☐ NO	11A. DO YOU REQUIRE GOVERNMENT CONTRACT FINANCING TO PERFORM THIS PROPOSED CONTRACT? *(If "Yes," complete Item 11B)* ☐ YES ☐ NO	11B. TYPE OF FINANCING *(√ one)* ☐ ADVANCE PAYMENTS ☐ PROGRESS PAYMENTS ☐ GUARANTEED LOANS
12. HAVE YOU BEEN AWARDED ANY CONTRACTS OR SUBCONTRACTS FOR THE SAME OR SIMILAR ITEMS WITHIN THE PAST 3 YEARS? *(If "Yes," identify item(s), customer(s) and contract number(s))* ☐ YES ☐ NO	13. IS THIS PROPOSAL CONSISTENT WITH YOUR ESTABLISHED ESTIMATING AND ACCOUNTING PRACTICES AND PROCEDURES AND FAR PART 31 COST PRINCIPLES? *(If "No," explain)* ☒ YES ☐ NO	

14. COST ACCOUNTING STANDARDS BOARD (CASB) DATA *(Public Law 91-379 as amended and FAR PART 30)*

A. WILL THIS CONTRACT ACTION BE SUBJECT TO CASB REGULATIONS? *(If "No," explain in proposal)* ☒ YES ☐ NO	B. HAVE YOU SUBMITTED A CASB DISCLOSURE STATEMENT *(CASB DS-1 or 2)? (If "Yes," specify in proposal the office to which submitted and if determined to be adequate)* ☒ YES ☐ NO
C. HAVE YOU BEEN NOTIFIED THAT YOU ARE OR MAY BE IN NON-COMPLIANCE WITH YOUR DISCLOSURE STATEMENT OR COST ACCOUNTING STANDARDS? *(If "Yes," explain in proposal)* ☐ YES ☒ NO	D. IS ANY ASPECT OF THIS PROPOSAL INCONSISTENT WITH YOUR DISCLOSED PRACTICES OR APPLICABLE COST ACCOUNTING STANDARDS? *(If "Yes," explain in proposal)* ☐ YES ☒ NO

This proposal is submitted in response to the RFP contract, modification, etc. in Item 1 and reflects our best estimates and/or actual costs as of this date.

15. NAME AND TITLE *(Type)*	16. NAME OF FIRM

17. SIGNATURE	18. DATE OF SUBMISSION

NSN 7540-01-142-9845

1411-101

☆ U.S. GOVERNMENT PRINTING OFFICE : 1984 O - 421-526 (37)

STANDARD FORM 1411 (10-83)
Prescribed by GSA
FAR (48 CFR) 53.215-2(c)

Figure 4-3 *Standard Form 1411*

DESCRIPTION	PAGE TYPE QUANTITIES						TOTAL PAGES	MEETINGS/ DELIVERABLES
	TEXT	TABLE	HALF-TONE	LINE DRAWING	FUNCT-IONAL	ENG DWG		
FRONT MATTER	12						12	DRAFT MANUSCRIPT ☐
CH 1 - GEN INFO	5	4		4	2		15	
CH 2 -INSTALLATION								
0 INSTAL LOG	1			2			3	REPRO/CAMERA READY COPY ☐
0 INSTAL PROCED	1						1	
CH 3 - PREP FOR USE								
0 PREP FOR USE	1	3		3			7	NEGATIVES ☐
0 PREP FOR RESHIP	1	1		1			3	
CH 4 - OPERATION								
0 CONT & INDICATOR	1	3		1			5	BOARD ART ☐
0 OPER INSTRUCT	5	10		4			19	
0 EMER OPER	1						1	
CH 5 - THEOR OF OPER								
0 FUNCT SYS OPER	9			5	5		19	VALIDATION ☐
0 FUNCT OPER - CKT CARDS	40			10	10		50	
CH 6 - MAINTENANCE								
0 ORG/INTERMED	5	10					15	VERIFICATION ☐
0 SPECIAL MAINT	50	75		12			137	
0 PERFORM TEST	4	10		1			15	IN-PROCESS REVIEWS
CH 7 - DIAGRAMS						60	60	☐ ☐ ☐
TOTALS	136	116		43	17	60	362	

START DATE _____
STOP DATE _____
CONTRACT AWD DATE _____
TECH WRITER _____

APPLICABLE SPECIFICATIONS:
_____
_____
_____
_____

COMMENTS:

Figure 4-4 *Typical Proposal Estimate Worksheet*

your page count numbers represent. You may have more difficulty deriving a detailed BOE from this type of estimate and, therefore, justifying your page counts to your management and to the customer.

The hardware breakdown method of estimation is risky and only those with extensive experience in bidding should attempt to use it. The page count evaluation process requires a skilled technical writer who has vast experience preparing manuals for many customers on many types of equipment, written to many different specifications. The writer must be thoroughly skilled in converting engineering data into text and illustrations.

After the page count estimate is completed, the identity and location of exactly where the pages of text, tables, and illustrations will be used in the manual is almost totally lost. Justification of page counts is extremely difficult. The writer will be forced to go back to the total page count number to get a rational explanation of the chapter by chapter breakdown of pages. Assigning labor hours to the various labor categories (writer, editor, illustrator, etc.) for the quantity of page types also becomes more difficult. The bidder must account for all factors impacting total hours per page, such as time for customer reviews and time to make changes due to both customer reviews and engineering design changes.

Estimating Labor Cost Per Hour

Several labor categories (types of labor, such as writer, editor, illustrator, etc.) are needed to write the technical manuals, to verify their accuracy, and to produce the docu-

ment from start to completion of camera-ready copy. Normally the categories include the following list. (Note: costs for the following labor categories may be offered in the proposal in any of several areas. This book assumes that the costs for all labor categories except Quality Control, QC, and safety are part of the technical publications costs.)

- Technical writers to write the data
- Word processors to produce the pages of text, tables and illustrations
- Illustrators to draw the illustrations, touch up photos, mount art, etc.
- Engineers to provide source information, to evaluate the technical accuracy of data prepared by the writer, and to participate in validation and verification exercises
- Editors/proofreaders to check all aspects of grammar and to make sure the material has been properly produced.
- Photo lab personnel to prepare stats, negatives, etc.
- Print shop personnel to print copies of the manuals
- QC and safety personnel to evaluate the final documents and to confirm that all necessary aspects of user safety are included

Estimating labor hours for each category is accomplished by deriving a BOE for work to be done. Ranges of time for work done per hour should be based on documented, provable statistics. For each Labor category within industry there are general standards which represent averages of efficiency. For example, a technical writer should be able to write one page of data in a given length of time. Writing covers all efforts involved, including planning, research, meetings, changes due to engineering modifications and customer reviews, the production process control, etc. A word processor should be able to enter (type) a standard page of data at a set rate of pages per hour. Similarly, average standards for generating various types of illustrations should be derived.

These averages should be based on documented analysis of previous tasks. If not documented, the averages fall into the realm of "engineering estimates." Within industry there are averages that can be used for estimating hours to perform the various tasks. The following are examples:

Technical Writer: 2–3 hours/page (slightly complex)
                      3–4 hours/page (more complex)
                      4–5 hours/page (very complex)

Illustrator: 2–5 hours/page (slightly complex)
                      5–10 hours/page (more complex)
                      10–25 hours/page (very complex)

These examples reflect the original preparation time plus the time to make changes due to In-Process Reviews (IPR's), validation and verification. Each page represents a typical 8-½ × 11 inch page. For each additional page or foldout (i.e. 11 × 17-inch piece of art), the time is doubled.

Estimating Cost of Materials

The materials costs fall under Other Direct Costs (ODC) related to technical publications. Examples of materials which have to be considered in the proposal cost include:

- Negatives (opaqued as necessary)
- Stats or other form of hardcopy used as original art for Camera-Ready Copy
- Materials (cardboard, tissue paper, draft paper, adhesive glue) for boardmounting art
- Covers and binders (heavy paper, loose-leaf, spiral, binding posts, fasteners, etc.)
- Magnetic media (disks, tape, etc.)

All costs for materials will vary depending upon where your company purchases them and the quantity of materials purchased. Writers involved in costing materials are dependent upon cost data supplied to them by company purchasing agents or the functional organization which handles these materials. Writers do not normally become involved in negotiations for purchasing materials. All materials costs related to deliverables and quantities of deliverables which are cited in the contract must be justifiable as part of the manual's BOE, if the government later decides to conduct a fact finding probe. As the technical writer involved in bidding, you must be able to determine types and quantities of materials. Some factors involved include:

- Deliverable Types
  Draft Manual
  Final Manual
  Camera-Ready Copy
  Boardmounted Art
  Negatives
  Printed (Hard) Copies

- Quantities
  Defined by CDR
  Defined in the SOW

A typical CDRL (Contract Data Requirements Lists) lists one or more addresses and quantities for each deliverable. Under certain circumstances, the SOW may stipulate a requirement for additional copies for other purposes, such as copies to be packed with each item of hardware and copies to support training. There may also be a stipulation that should the contractor be unable to complete final manuals before delivery of hardware, the contractor must pack advance copies of the manual and replace them later when the final manual is completed. The contractor is also responsible for providing the customer with copies of material prepared for IPR's. The number of copies should be sufficient for all attendees (customer and contractor) to the IPR's, including house copies used for reviews by engineers. If the manuals are to be validated, copies must be made for each participant. Copies may also be required to support a maintainability demonstration.

If your company has its own print shop and photo lab, internal costs are used. If it does not, costs (which must be documented) must be obtained from an outside source. Deriving total materials costs becomes a matter of multiplying quantities by types of materials and adding up the total.

Manual Costing Considerations

After estimating the number of pages, the time to produce each page by the different types of labor, and estimating the cost of materials, you can cost the manual. However, you must be certain all factors have been considered. Writers need time to do research, and this time is not directly measurable by hours. Learning curve time is also required because writers do not automatically understand the equipment.

As discussed previously, determining the accurate number of pages needed in a manual plus the extra hours and costs needed to prepare those pages takes experience. The following are but a few questions you must ask if you plan to produce an acceptable proposal cost package which identifies all Direct Labor (DL) costs and Other Direct Costs or ODC.

1. Equipment Complexity

a. How many racks, chassis, circuit cards, power supplies (Line Replaceable Units or LRU's) are there?

b. How dense are the circuit cards? (How many piece parts are there, and how many and how complex are the digital chips or DIP's?)

c. What is the maintenance concept required by the contract? (Are the users of the hardware designated operator, organizational unit, field or intermediate, off-site, or depot level maintenance? What level of repairability has been defined? Are any LRU's throw-away, rather than repairable?)

d. How will the units/chassis/racks be tested? (Will Built-in-Test (BIT), or Built-in-Test Equipment, (BITE), be available or will everything have to be tested manually?)

e. How many unique types of LRU's will there be?

2. Specification Requirements

a. Are the manuals to be prepared to military or commercial specifications?

b. What is the level of the manual required by the contract? (a direct-tie-in with the maintenance concept: are we writing for an operator, or one of the maintenance echelons, or combinations of two or more?)

3. Data Availability

a. What types of source data will be available? Will Level 1, 2, or 3 engineering drawings be available for your illustration requirements? Will you have a CDR package or another similar source data package from which to obtain information?

b. Will your engineering personnel be preparing rough engineering level text for you or will your technical writers be preparing most of the data based upon discussions with the engineers and review of their design material?

c. Will you have other sources of information available to assist in preparation of deliverables, such as Logistics Support Analysis (LSA), LSA Record (LSAR), Parts Location Diagrams prepared for other manuals, Illustrated Parts Breakdowns (IPB's), Repair Parts and Special Tool Lists (RPSTL's), or other manuals from which you will extract material?

d. Will your time schedule permit waiting until information is more or less ''cast in concrete,'' or will you be required to start writing while the engineering design is still fluid and likely to change as you are writing?

e. When will the hardware be available for you to refer to while preparing your information, such as parts removal and replacement procedures?

4. Contract Requirements

a. Will you be required to have IPR's?

b. Will you have a validation or verification or both?

c. Will you have to travel?

d. Do you have to prepare any other preliminary data, such as a technical manual plan, outlines, maintenance allocation chart or support equipment recommendations, which will be drivers for your manual's content?

e. Is your schedule for deliverable material realistic? Can you meet the deadline with your existing staff?

5. Customer Requirements

a. Are you familiar with your customers? Do you know their requirements and interpretation of both the designated specifications and contract requirements?

b. Is your customer dependent upon your schedule as part of a larger schedule where your delivery dates lie in the critical path of some other requirement?

Types of Contracts

After receiving the various proposals from contractors, the customer evaluates the offers. The contract is awarded on the basis of the contractor's technical competence, cost, ability to perform, compliance to the requirements cited in the bid package, and understanding of the contract. Depending upon the type of contract, cost alone may be the key factor in the decision as to which contractor wins. The selected winner is notified and the contract start date is set. This date becomes the After Date of Award Document (ADAD) or After Receipt of Order (ARO) date. It is the date to which CDRL items are keyed.

There are several major types of contracts used by the government. The following is a list of seven with their definitions and reasons for their selection:

1. Firm Fixed Price (FFP)	Used when the equipment design has reached a stage of reasonable development. Contractors can usually make reasonable cost estimates.
2. Fixed Price with Economic Price Adjustment (FP-EPA)	Used when conditions such as market or labor are unstable, and identifiable contingencies can be separately priced.
3. Fixed Price Incentive (FPI)	Used when there are uncertainties in cost, and where either improvements in performance or cost reduction may exist. A performance incentive fee may be offered.
4. Fixed Price with Redetermination (FPR)	Is either prospective (used when a reasonable price can be negotiated for at least part of the contract) or retroactive (used when a reasonable fixed price cannot be negotiated at the start of the contract).
5. Cost Plus Award Fee (CPAF)	Used when performance cannot be accurately measured, but the contract can be completed and award fees are desirable incentives.
6. Cost Plus Incentive Fee (CPIF)	Used when hardware can be developed with a reasonable degree of certainty and performance incentives can be negotiated.
7. Cost Plus Fixed Fee (CPFF)	May be used when the task effort is not clear, and performance by the contractor is subject to interpretation. The contractor receives a fixed fee no matter what performance costs are incurred.

Conclusion

Various types of contracts are used by the government to meet different needs in the procurement process. The RFP, RFQ, and IFB are typical types of solicitations which contractors respond to when submitting proposals. After reviewing the proposals the government issues a contract to the winning contractors. While technical writers are not usually involved in direct negotiations, it is beneficial for the writers to understand both the content and the ramifications of the contract for which they are preparing documentation. They may be directly involved in development of technical data and cost inputs to the technical publications portion of the proposal. To perform their part of the task competently, they must be fully knowledgeable in publications requirements and experienced in documentation preparation. Interpreting specifications and estimating the size and content of technical documents should be done, however, with care.

Chapter 5

Re-Evaluating the Sources Before Writing the Manual— Planning the Job

Introduction

The time it takes after the proposal is submitted to the customer for the contractor to be notified of contract award can vary anywhere from weeks to a year or more. During this period, the customer and the program managers may have renegotiated the final specifications, Statement of Work (SOW), contract data requirements and delivery schedule. As a result of the negotiations, the work you planned and described in the proposal may be very different from the work you are now expected to do. Customer needs and/or equipment designs may have changed. You may also have to establish an internal agreement with the Integrated Logistics Support Manager (ILSM) or program manager based on the awarded contract and make a detailed plan which assures delivery "on time" and within cost goals.

Kick-Off Guidance Conference

The Kick-Off Guidance Conference (or Technical Publications Guidance Conference) is typically held within 30–90 days after the contract is awarded and serves several purposes. First, the meeting allows the parties to meet and establish one-on-one contact. Second, it provides the earliest possible chance for the parties to discuss how each perceives or interprets the contract requirements.

Even though there are specifications cited in the contract and even though the proposal describes the contractor's interpretation of the contract requirements, there is still room for misunderstanding and differences of opinion. The buyers (customers) think they know what they bought, and the sellers (contractors) think they know what they sold. But, until both parties actually discuss the contract, ask questions, and agree on all the major and minor points of the contract, neither can be certain that they understand perfectly how the other views the contract.

The particular specification cited in the contract or the SOW may be unclear. Points requiring clarification can be both major and minor. The goal of the conference is to get both parties to agree on what is required to fulfill the contract. All this is done so the customers will get their product on schedule and within cost and the contractors will get full renumeration for work done.

Those attending the conference include the technical publications personnel from both organizations and can also include Integrated Logistics Support Management (ILSM)

personnel as well as program management and contracts representatives. The technical publications person may represent only one area of discipline (technical manuals for hardware) or several disciplines (hardware, software, and parts listing personnel). Or, each discipline may be separately represented by different individuals. If the program is Tri-service (Army, Navy, and Air Force), representatives from all three government branches may be present. The usual practice is for one branch to be the host. Other interested parties may or may not wish to be active participants in the technical manual development process. Participation depends on involvement and interest in the program. Those having a large financial stake and those who will be users of the manuals will be more involved in the decision making. At a conference table, it is not unusual for a small group of two or three contractor representatives to face a force of ten or more governmental representatives.

Technical Writers at the Conference

Technical writers attending the conference should possess sufficient skill to understand their responsibilities and limits of authority given them by the company. They are there to ask questions and to provide answers; they must know when to accede to customer demands, when to tactfully stand their ground, and when to recognize the limits of their authority. At some point it may be wise to say, "I'll have to check into that and let you know at a later date."

Resolving Conflicts at the Conference

There may be times when the customer and contractor disagree either during or after the Kick-Off Conference. Whenever a problem arises related to contract interpretation which cannot be resolved by those attending the conference, the first recourse is to inform upper level management, in other words "to escalate." Bear in mind, the technical publications effort for a military procurement is usually only a small part of the overall contract and should not jeopardize the whole contract.

Re-Evaluating the Proposal

Re-evaluating the sources before writing the manual includes analyzing the original proposal and the contract. Each is explained in more detail in the next paragraphs. The proposal's Technical section defines the technical pub-

CONTRACT DATA REQUIREMENTS LIST

ATCH NR 2 TO EXHIBIT E

TO CONTRACT/PR _____

CATEGORY ____ M

SYSTEM/ITEM MODEM AN/XXXX

CONTRACTOR XYZ CORP

1. SEQUENCE NUMBER	2. TITLE OR DESCRIPTION OF DATA / 3. SUBTITLE / 4. AUTHORITY (Data Item Number) / 5. CONTRACT REFERENCE	6. TECHNICAL OFFICE / 7. DD250 REQ / 8. APP CODE (A) / 9. INPUT TO IAC (X)	10. FREQUENCY / 11. AS OF DATE	12. DATE OF 1ST SUBMISSION / 13. DATE OF SUBSEQUENT SUBM/EVENT ID	14. DISTRIBUTION AND ADDRESSEES (Addressee — Regular Copies/Repro Copies) / 15.
E002	2. Technical Manuals/Commercial Literature DEPTM 11-( )-( )-14 4. DI-M-6153 16. REMARKS 1. Prepare in accordance with Supplement A to Data Item DI-M-6153 and Statement of Work — Narrative Manuals. After approval of review ms/art, printed copies will be submitted.	6. W15P66 7. DD 8. A 9.	10. OTIME 11.	12. 460 days ADAD re- 13. view MS 510 days ADAD: After printed copies	14. W15P7B DRSEL-ME-PCN Manuscript 7/0 Artwork 0/1 Printed Cys 50/1 15. TOTAL 57/2
E003	2. Technical Manual Plan (TMP) 3. Technical Manual Outline 4. DI-M-6154 16. REMARKS 1. Prepare in accordance with block 10f, Data Item DI-M-6154, Supplement A to DI-M-6154, and Statement of Work — Narrative Manuals.	6. W15P66 7. DD 8. A 9.	10. OTIME 11.	12. 100 days ADAD 13.	14. W15P7B DRSEL-ME-PCN 5/0 DRCPM_SC-8B 3/0 15. TOTAL 8/0
E004	2. Technical Manual Plan (TMP) 3. Validation Schedule 4. DI-M-6154 16. REMARKS 1. Prepare in accordance with block 10h, DI-M-6154, Supplement A to DI-M-6154, and Statement of Work- Narrative Manuals.	6. W15P66 7. DD 8. A 9.	10. ASREQD 11.	12. 300 days ADAD 13.	14. W15P7B DRSEL-ME-PCN 3/0 DRCPM-SC-8B 3/0 15. TOTAL 6/0
E005	2. Validation Record (Technical Manuals) 4. DI-M-6159 16. REMARKS 1. Prepare in accordance with DI-M-6159, and Statement of Work — Narrative Manuals. 2. Submit concurrent with submission of publication manuscript	6. W15P66 7. DD 8. 9.	10. ASREQD 11.	12. See block 16 13.	14. W15P7B DRSEL-ME-PCN 3/0 15. TOTAL 3/0

PREPARED BY	DATE	APPROVED BY	DATE	PAGE ___ OF ___ PAGES

DD FORM 1423 1 JUN 60   REPLACES EDITION OF 1 APR 68, WHICH IS OBSOLETE.

Figure 5-1 *Typical Contract Data Requirements List—DD Form 1423*

INSTRUCTIONS FOR COMPLETING DD FORM 1423

For Government Personnel

1. This form (or its equivalent adapted for ADPE) shall be used whenever data is required to be delivered under a contract. The form (except Items 23 through 26) shall be completed in accordance with Departmental procedures, and furnished to the contracting officer by the personnel responsible for determining the data requirements of the contract.

2. This form shall constitute the sole contractual list of requirements for the amounts and kinds of data to be required by the contract (but see paragraph 1 below).

3. Local reproduction of this form in exact format, 13'' x 8'', is authorized.

For the Contractor

1. OMISSIONS FROM DD FORM 1423. Failure of the Government to list on this form any item of data which is required to be delivered, or whose delivery the Government has the right to require, under any clause included in this contract shall not relieve the Contractor of his obligation under such clause.

2. The Contractor agrees that, regardless of whether he has made any entries in Items 25 and 26, and regardless of what those entries are, he is obligated to deliver all the data listed hereon, and the price he is to be paid therefor is included in the total price specified in this contract.

3. The estimated prices filled in in Item 26 will not be separately used in evaluation of offers.

4. Each offeror may complete Items 23 and 24 in accordance with the following instructions:

Item 23. Contractor File/Document Number - Enter bidder's or offeror's internal filing or document number, if applicable.

Item 24. Estimated Number of Pages - Enter the estimated number of pages, drawings, etc. for single preparation.

5. Each offeror shall complete Items 25 and 26 in accordance with the following instructions (this does not apply to advertised contracts or to negotiated contracts under $100,000).

Item 25. Price Group - Contractors shall specify one of the four following groups of effort in developing estimated prices for each item of data listed on the DD Form 1423.

a. Group I. Definition - Data which is not otherwise essential to the contractor's performance of the primary contracted effort (production, development, testing, and administration) but which is required by DD Form 1423.

Estimated Price - Costs to be considered under Group I are those applicable to preparing and assembling the data item in conformance with Government requirements, and the administrative and other expenses related to reproducing and delivering such data items to the Government.

Example for Group I - A technical manual prepared for military use only. The estimated price of the manual would be noted on the DD Form 1423 exclusive of costs for any of the manual material that had been generated for other purposes (e.g., drawings used both for production and as illustrations in the manual).

b. Group II. Definition - Data which is essential to the performance of the primary contracted effort but which the contractor is required to perform additional work to conform to Government requirements with regard to depth of content, format, frequency of submittal, preparation, control or quality of the data item.

Estimated Price - Costs to be considered under Group II are those incurred over and above the cost of the essential data item without conforming to Government requirements, and the administrative and other expenses related to reproducing and delivering such data item to the Government.

Example for Group II - In the case of MIL-D-1000 Form I drawings (drawings to military standards), the estimated price of the data item begins only after the engineering and manufacturing information has been initiated. The estimated price shall not include the cost of configuration control but shall include any additional quality assurance and control of the drawings but not related to engineering configuration control. Not to be considered is "design effort" expended on layout drawings and other data which serve principally as a medium for developing design and are not used in manufacture, production or test of the end item.

c. Group III. Definition - Data which the contractor must develop for his internal use in performance of the primary contracted effort and does not require any substantial change to conform to Government requirements with regard to depth of content, format, frequency of submittal, preparation, control and quality of data.

Estimated Price - Costs to be considered under Group III are the administrative and other expenses related to reproducing and delivering such data items to the Government.

Example for Group III - A drawing prepared to Form 2 or 3 of MIL-D-1000 (drawings to company standards) which had been used in the manufacturer's normal plant activities.

d. Group IV. Definition - Data which is developed by the contractor as part of his normal operating procedures and his effort in supplying these data to the Government is minimal.

Estimated Price - Group IV items should normally be shown on the DD Form 1423 at no cost.

Example for Group IV - A brochure or short manual used in a company's normal commercial business, that is acquired by the Government in such small quantities that cost of determining a charge would not be practical.

Item 26. Estimated Total Price.

a. For each item of data listed the bidder or offeror shall enter an amount equal to that portion of the total price which is estimated to be attributable to the production or development for the Government of that item of data. These estimated data prices shall be developed only from those costs which will be incurred as a direct result of the requirement to supply the data, over and above those costs which would otherwise be incurred in performance of the contract if no data were required.

b. The estimated data prices shall not include any amount for rights in data. The Government's right to use the data shall be governed by the pertinent provisions of the contract.

GPO 801-250

Figure 5-1 Continued

DATA ITEM DESCRIPTION	2 IDENTIFICATION NO(S)	
	AGENCY	NUMBER
1 TITLE Technical Manual Plan (TMP)	DOD	DI-M-6154

3 DESCRIPTION/PURPOSE
To prescribe the general procedures, terms, and conditions governing the planning, selection, preparation, and delivery of technical manuals required for operation and maintenance of aeronautical system/equipment being procured.

4 APPROVAL DATE
30 April 1971

5 OFFICE OF PRIMARY RESPONSIBILITY
NAVY.

6 DDC REQUIRED

6 APPROVAL LIMITATION

7 APPLICATION INTERRELATIONSHIP

This data item is to be used during acquisition of technical manual programs supporting multiservice aeronautical systems. The contents of the plan are flexible, and the extent of coverage required may be decreased dependent upon the size or complexity of the contracts.

DI-M-6153 Technical Manuals/Commercial Literature
DI-M-6155 Technical Manual Status and Schedules

This plan is the Technical Manual Section of the Integrated Support Plan (ISP) required by Part 2, Section 1, DI-L-6138 of the SISMS Contract and Data Requirements.

9 REFERENCES (Mandatory as cited in block 10)

AMCP 700-4
NAVMAT-P-4000-1
AFLCM/AFSCM 400-4
MCO P4110.1

MCSL NUMBER(S)

10 PREPARATION INSTRUCTIONS

The contractor shall provide a TMP clearly defining the intended purpose of each manual, delineating the scope of each manual, and explaining the interfaces and overlap between or among the manuals. The TMP shall include, but shall not be limited to, the following items of information:

a. <u>Description of General Plan for Evolving Technical Manuals</u> - A complete analysis of the technical manual program for the particular system or equipment under consideration. This shall include a summary of any assumptions, conditions, or limitations affecting the overall TMP along with the reasoning and specific impact of each on the plan.

b. <u>Method of Use of Data</u> - A detail description of procedures which will assure that all pertinent design, operation, and maintenance data is adequately, accurately, and clearly reflected in the applicable technical manuals.

c. <u>Method for Achieving Standardization</u> - A detailed account of procedures to achieve standardization in writing style, art work, nomenclature, abbreviations, and symbols.

d. <u>Use of Specifications</u> - A list of all specifications applicable to the technical manual programs. Any problems regarding specification interpretations and applications shall be identified in this section.

e. <u>Integration and Coordination Between Contractors</u> - An explanation of the methods to be used to relate and control the integrating and associate/subcontractors' (cont)

DD FORM 1664

Figure 5-2 Typical DID (DI-M-6154) DD Form 1664

lications tasks. The manuals needed usually are identified in the form of a manual tree, along with description of the format, content, tables, illustrations, and size of the manual. The maintenance concept and operation levels may also be detailed. The information originally used in the proposal to estimate the cost of the manual must be re-evaluated in light of new costs (which now could make your original estimates inaccurate) and data requirements.

The goal of any government contract is to obtain a product that fills a need. But the need defined by a Request For Proposal (RFP) or Request For Quote (RFQ) may not be the same need drawn up in the final contract. Customers may alter their design concept after the contract award. Also contractors may decide to alter the approach to the design for the benefit of the customer and themselves. Or after evaluation, the contractor may decide to make or buy hardware which shifts all or part of the need for data from the contractor to a subcontractor. During the Technical Publications Kick-Off Conference, the customer may waive certain requirements or trade off one requirement for another.

Any one of these changes can affect the original cost and schedule of the technical publications (the manual). If commercial off-the-Shelf (COTS) manuals are included in the deliverable equipment or if a subcontractor will handle part of the technical publications, the number of pages the contractor's technical publications people will have to prepare decreases. When the number of pages decreases, the cost decreases for the contractor's technical publications people. Also, if the number of remaining pages to be written decreases, the number of in-house writers required to perform the task also decreases. Part III of this book provides more detailed information on technical publications subcontracting.

Analyzing the Contract

The contract usually contains a few key areas which define contractor requirements for technical data. These areas include the Contract Data Requirements Lists (CDRL's), Data Item Descriptions (DID's), Attachments and Addendums to DID's, and the SOW. Figures 5-1, 5-2, 5-3, and 5-4 provide typical illustrations of these four items from which the technical writer obtains details such as what to write, what format to use, what control methods to use, and how many of each deliverable items are required. The CDRL, DID and SOW are discussed in more detail in the following paragraphs.

● Analyzing the CDRL's

Customers use the CDRL to stipulate what data is required from the contractors. The typical CDRL is prepared on DD Form 1423 (Figure 5-1). It is divided into numerous blocks which are numbered. In this sample there are fifteen duplicate boxes, one containing a title and one a blank for each number. For example, box 1 at the top left has the title SEQUENCE NUMBER. Two blocks below is another box numbered 1 which is blank. Customers enter the se-

quence number of this particular CDRL and enter the data which describes or defines their requirements into all the other empty boxes. Table 5-1 lists the blocks and defines their purpose.

● Analyzing DID's

The DID is used by the government to describe a particular item of data which it wishes to buy. The DID is prepared on DD Form 1664.

DID Functional Categories

Each DID category is identified as part of the Data Item (DI) identifying number. Table 5-2 lists the various categories under which data may be procured and briefly defines their content.

DID Numbering System

The DOD Acquisition Management Systems and Data Requirements Control List (AMSDL) issued by the Department of Defense (DOD) states "those source documents and related Data Item Descriptions (DIDs) which have been cleared for use in defense contracts by the Office of Management and Budget (OMB) under the authority granted by Public Law 96-511." DID's are numbered using four descriptors. The first indicates the requirements for a Data Item. The next descriptor may be one or more letters. Currently, DIDs use a four digit alpha symbol group to identify functional categories. Technical Manual Specifications and Standards (TMSS) covers technical content and format of weapon systems. Prior to July 1985 a single alpha character represented the functional category. In the example the letter identifies the category as selected from the list in Table 5-2. A series of digits follow, giving the DI's control number. The last descriptor, when used, indicates the revision of the DI number (i.e. DI-M-6154), DI, category M-technical publications, description number 6154.

This particular DID is for a Technical Manual Plan (TMP), a document the contractor prepares at the beginning of the program. It defines the contractor's plans for creating the manual, use of data which will be available to prepare the manual, what methods will be used to achieve standardization, which specifications will be used, and how coordination between contractors (when applicable) will be established, etc. Figure 5-2 contains the complete requirements for this DID.

● Analyzing the SOW

The SOW for Technical Publications in a government contract can range from being very brief to being very detailed. It sometimes just makes a few clarifying statements, and the government depends upon the DID's, CDRL's, and various addendums to provide most of the detailed requirements. Figure 5-4 is an example of a brief SOW. It identifies requirements about delivery of camera-ready copy for manuals, outlines, plans, and records and identifies the need for a Literature Guidance Conference

SUPPLEMENT A TO DATA ITEM DI-M-6153 (Narrative Manuals)

In accordance with the requirements, quantities, and schedules set forth in the Contract Data Requirements List (CDRL), DO Form 1423, Category M, and Supplement A thereto, the contractor shall perform as follows.

1. Literature Specifications. The following literature specifications including the latest issue at time of contract award apply:

MIL-M-38784	Manuals, Technical: General Style and Format Requirements
MIL-M-63036 (TM)	Manuals, Technical: Operator's Preparation of
MIL-M-63038 (TM)	Manuals, Technical: Organizational or Aviation Unit, Direct Support or Aviation Intermediate and General Support Maintenance
MIL-M-63007 (TM)	Manual, Hand Receipt (DA Form 2062)

2. Literature Requirements. The material for narrative publication DEPTM 11-( )-( )-14 Operator, Organizational, and Direct Support Maintenance Manual. Publication number shall be assigned by the procuring activity (DRSEL-ME-PCN).

a. The DEPTM manuscript shall be prepared in accordance with MIL-M-63036 (TM) and paragraph 3.5., MIL-M-63038 (TM).

b. The DEPTM manuscript shall be prepared in camera-ready copy, black and white, with single spacing, double column (para 3.2.2.2. MIL-M-38784), with page size 8 1/4″ × 10 3/4″ (para 3.2.2.2. MIL-M-38784), and other camera-ready copy requirements of MIL-M-38784.

c. The DEPTM shall be consonant with the maintenance allocation chart and with the tests, alignment, and troubleshooting procedures developed under Ground Support Equipment Recommendation Data (GSERD) cited elsewhere in this contract (Data Item DI-S-6176).

d. The material for appendices B (Components of End Item (COEI) and C (Maintenance Allocation Chart (MAC) shall be derived from other data items in the contract. The requirements to include appendices for Additional Authorization List (AAL) items and Expendable Supplies and Materials List (ESML) shall be determined during development.

e. Principles of Operation, paragraph 3.2.3.3. MIL-M-63038 (TM).

(1) For organizational maintenance, the principles of operation shall be limited to those which would enable the repairperson to test, troubleshoot, and repair the equipment to the parts/circuits identified in the maintenance allocation chart for organizational level.

(2) For direct/general support maintenance, the principles of operation shall cover the entire equipment including schematic diagrams of all circuits. These principles of operation shall precede the maintenance instructions (para. 3.2.4.6.) Maintenance Procedures, MIL-M-63038 (TM).

f. The manuscript shall be validated before submission in accordance with validation requirements specified in MIL-M-63038 (TM).

g. After approval of the review manuscript, the DEPTM shall be reproduced (printed) (para 5 below).

h. A technical manual outline shall be submitted in accordance with CDRL, DD Form 1423, Data Item DI-M-6154, section 10f, cited elsewhere in the contract.

3. DEPTM 11-( )-( ) 14-HR, Hand Receipt Manual shall be prepared in accordance with MIL-M-63007 (TM) and the following.

a. The Introduction, Section I, shall be prepared camera-ready with double spacing, single column across the page, with page size 8 1/4″ × 10 3/4″.

b. The procuring agency shall provide blank form DA Form 2062 to be filled out from the information in the Components of End Item List (COEIL) provided in the basic DEPTM.

c. The requirements to provide the Additional Authorization List (AAL) will be determined during the development of the DEPTM.

d. After submission and approval of the review manuscript, the DEPTM shall be reproduced (printed) (para 5 below).

4. DEPTM Changes. Equipment resulting from DT/OT-II that affect information in the printed DEPTM (not to include the Hand Receipt DEPTM) shall be covered by loose-leaf, camera-ready change to the DEPTM.

a. If the change material results in run-over pages/paragraphs, use point-numbered pages/paragraphs (para 3.7.2.2. MIL-M-38784). The procuring activity will supply the "master" DEPTM copy and illustrations.

b. DEPTM Change cover shall be prepared per figure 38, MIL-M-38784.

c. DEPTM Change manuscript shall be validated before submission for review and approval. Approved Change manuscript shall be reproduced (printed) (para 5 below).

5. Review Manuscript Copies and Printed Copies. These copies shall be produced from the "master" manuscript copy using good commercial reproducing practices (3M, XEROX, etc.).

6. Residual Material. Residual material, such as illustrations artwork and the "master" manuscript copy shall be delivered concurrent with the printed DEPTM copies.

Figure 5-3 *Typical DID Supplement*

STATEMENT OF WORK
NARRATIVE MANUALS

In accordance with requirements, quantities, and schedules set forth in Contract Data Requirements List (CDRL), DD Form 1423, Category M, the contractor shall perform as follows.

1. Submit camera-ready DEPTM 11- ( )-( )-14 in one color, 8-1/4″ × 10-3/4″ size, in accordance with CDRL, Data Item DI-M-6153, for Operator, Organizational, Direct Support and General Support Maintenance Manual.

2. Submit camera-ready DEPTM 11-( )-( )-14 HR in 6 1/4″ × 10-3/4″ size, in accordance with CDRL, Data Item DI-M-6153 for Hand Receipt Manual.

3. Submit Technical Manual Outline and Validation Plan/Schedule in accordance with Data Item DI-M-6154, Technical Manual Plan (TMP), blocks 10F, and 10h, respectively.

4. Submit Validation Records for each validation effort for above DEPTMs in accordance with CDRL, Data Item DI-M-6159.

5. A Literature Guidance Conference shall be held at Fort Monmouth, New Jersey between the contractor and the Government Procuring Activity within 45 days after contract award (and subcontractor, if applicable) for the purpose of providing guidance on the publication literature requirements and to formalize validation schedules and in-process reviews. Full advantage will be taken by the Government to perform in-process reviews to inspect the progress of the publication preparation in order to avoid delay in submission and equipment delivery.

Figure 5-4   *Typical Statement of Work*

(Kick-Off or Publications Conference). In this particular case, all other details related to the technical manual procurement are given in the other contract data.

An experienced person who looks at Figure 5-4 would immediately recognize that the SOW is from an Army contract for technical data. How? There are three clues.

The first clue is the reference to DEPTM 11-( )-( )-14. This technical manual number (only partially complete) is from the numbering system unique to the Army. The Army uses DEPTM to indicate Draft Equipment Publications Technical Manuals, while the Air Force uses TO to indicate Technical Orders.

The second clue is -14 type manual which is one that contains operation and maintenance information for all levels from Operator through General Support (GS) level. It is one of a series of numbered manuals identified in the Army's MIL-M-63036 and MIL-M-63038 Military Specifications for technical manuals.

The third clue is the Literature Guidance Conference in Fort Monmouth, New Jersey, which is the location of the Army's CECOM/SATCOMA technical publications center. This information is a key because it indicates that even if the requirement were tri-service (Army, Navy, Air Force), it takes place at the Army's installation and points to the Army as the prime contract controlling agency.

On some contracts, military service branches which will be users of the acquired hardware and manuals but are not in control of the contract may still require data unique to their branches. In other words, a contract where the Air Force is the user (but is not in control) may contain CDRL requirements for a Maintenance Allocation Chart (MAC). The MAC is an Army document used to identify repairable assemblies, types of maintenance, tools required for maintenance, and times required to perform maintenance. One critical area of difference in service branches is the maintenance concept. While one may find another's technical manual usable, it may still have difficulty provisioning spare parts or identifying various maintenance levels for its personnel.

A technical writer's understanding of the SOW is central to an overall understanding of the technical publications and the customer's specific needs. Technical writers must be extremely careful when reading contractual requirements. Failure to read every word carefully or to make assumptions hastily can have a disastrous impact on their ability to meet contract requirements.

Planning Ahead

After the proposal and contract have been re-evaluated, the technical writer now must plan ahead by coming up with a definite program schedule. A program schedule is usually included in the original proposal. This schedule identified major milestones such as the Preliminary Design Review (PDR), Critical Design Review (CDR), Acceptance Test Procedures (ATP), First Article Delivery, etc. The schedule showed how the technical publications requirements fit into the overall program schedule. The schedule was developed from the contractual dates identified in the CDRL's, addendums to DID's, and the SOW. Figure 5-5 illustrates a typical program schedule with all the key milestones.

Technical Publications Timetable

Technical writers usually need to create a more detailed schedule after the contract has been won to show more frequent inspections or milestones. If they know the start and end dates, they can set up a baseline schedule. The contract identifies other key milestones the customer believes to be important. These milestones include conferences, In-Process Reviews (IPR's), dates for draft and final versions of the various deliverables, validation and verification, and possibly changes caused by modifications to the hardware. Also included are requirements to incorporate changes resulting from errors detected during training or other contractual testing such as maintainability, operational testing or follow-on evaluations.

A schedule to identify the milestones for technical pub-

Table 5-1 *CDRL Format*

Block	Title	Description
1	Sequence Number	An assigned number to delineate CDRL as defined by ASPR regulations
2	Title of Description or Data	DD Form 1664 title
3	Subtitle	Additional descriptive data if title is not sufficient
4	Authority	Data Item Description number on DD Form 1664
5	Contract Reference	Paragraph number from contract which relates the CDRL to the contract and provides further clarifying information
6	Technical Office	Identification of the government office which has primary control and responsibility for the task
7	DD 250 Req	DD Form 250 (Gov Inspection/ Acceptance required; codes are SS Source/Source DD Destination/Destination SD Source/Destination DS Destination/Source XX Inspection/Acceptance defined in contract
8	APP Code	"A" indicates written approval required by Technical Office
9	Input to IAC	"X" indicates data is to be supplied to Integrating Associate Contractor
10	Frequency	Identifies frequency deliverable item is to be delivered: OTIME one time· One/R One time with revisions R/ASR Revisions as required ONE/P One time, preliminary draft ASREQ As required Daily Daily Weekly Weekly Bi-We Every two weeks Mthly Monthly Bi-Mo Every two months Qtly Quarterly (3 months) Semia Semiannually (6 months)
11	As of Date	Effective due date (calendar days) for CDRL at frequency defined in Block 10.
12	Date of 1st Submission	Identifies date for first submission of CDRL. If additional space is required for an expanded definition, refer to Block 16.
13	Date of Subsequent Subm/Event	Dates for additional submittals, when required
14	Distribution and Addresses	Identifies coded addresses and distribution quantities of deliverable CDRLs.
15	Total	Identifies total quantities of deliverable CDRLs cited in Block 14. Quantity types (copies and originals) are indicated by a slash: e.g., 5/1 means five copies and one original.
16	Remarks	Clarifying remarks are included to further define requirements.

lications will have time spread across the top and separate entries down the page for each task. Various symbols (i.e. a triangle) can be used to indicate the start, intervening events, and end of each task. Definite sequences are hinged to other events of the technical publications tasks or to the overall major program milestones. For example, the first event is typically the Guidance or Kick-Off Conference. This conference may take place within 30 to 90 days after contract award (ADAD or DAC). If technical publications tasks are identified in the development of "early deliverables" such as the Support Equipment Recommendation Data (Air Force) or Ground Support Recommendation Data (Army) or Maintenance Allocation Chart (Army), these tasks will occur soon after the conference is held. The actual technical manual writing effort starts when enough engineering data is available to the technical writer. This start time is typically at the CDR, when engineering documentation is officially released.

Technical publications tasks can be broken up into several milestones. Usually the number of IPR's depends upon the customer and the contract. Each IPR requires a certain percentage of the completion of the overall task to be available for review. If a task takes one year to complete and two IPR's are scheduled at the 30 percent and 70 percent com-

pletion points and the manual has 1000 estimated pages, a technical writer can use this data to come up with more specific publication milestone breakdowns. It is difficult to pinpoint exactly how much you will have completed at a certain date. You are never 100 percent certain that the original page count estimate is correct. You may be 10 percent, 20 percent or even 30 percent off (high or low), which can affect the actual timetable of the manual.

You cannot write until you have collected information and sorted the data into packages that match the areas of the manual you plan to write. You begin writing that part of the manual for which sufficient data is available. The lack of data causes the writing progress to be unlinear. For the first month or two it will be an almost flat line. Only later will you be able to start writing efficiently. Once the task is truly under way and more engineering data is available, the line will become linear-like. The line can be converted into an approximate number of pages written each month. This becomes a guideline for measuring your progress.

Milestone Chart

You may need to prepare a milestone chart. You enter time across the top, from ADAD (date contract is awarded)

Table 5-2 *DID Functional Category Descriptions*

Category	Title	Description
A	Administrative Management	Data required to administer and manage contractual requirements
E	Engineering/ Configuration data	Drawings, specifications, design data
F	Financial	Data related to forecasts, cost expenditures, etc.
H	Human Factors	Data related to human engineering safety and training
L	Logistic Support	Data on supply, general maintenance plans and reports, transportation and handling, preservation and packaging
M	Technical Publications	Data related to technical manuals on Installation, Operation and Maintenance
P	Procurement/ Production	Data related to both procurement and reprocurement, production and management information, including reporting, scheduling, facilities, etc.
R	Related Design Requirements	Data related to reliability, maintainability, survivability, etc.
S	System/Subsystem Analysis	Data related to technical reporting, modeling, performance related development, systems engineering
T	Test	Data related to plans and procedures and results concerning systems, equipment and part testing.
V	Provisioning	Data related to support of the provisioning process for both preoperational and operational support, including prescreening (performing a determination of current listing status), cataloging.

out through the very end of the technical manual tasks (through delivery of finals, negatives, and possible changes). Figure 5-6 is a typical example of a technical publication's milestone chart. You build the chart by adding key publication tasks in proper time relation to program events as shown in figure 5-5. You establish a key of consistent symbols and identify program milestones, showing the relationship between them and the start of the technical publications task. You identify start date, IPR's, draft and final deliverables, validation, verification and end dates. When comments are expected from reviews and training courses, these incremental milestones can then be added which are not tied to contract dates or to program milestones. You now have completed a milestone chart that will serve as the time budget for completing the writing tasks.

Personnel Projections

When the proposal was prepared, publications tasks were identified and hours-to-perform were estimated. If the hours are still accurate, you can set up a personnel projection chart. Figure 5-7 illustrates a typical projection chart. It identifies the types of personnel required to prepare or assist in the preparation of the technical publications. The projection determines how many people will be needed and how much time will be required (in monthly intervals) to prepare the manual. The projections relate labor categories to specific tasks. The government has standard labor categories for technical publications. Table 5-3 identifies the types of personnel and their functions.

You can also spread Other Direct Costs (ODC) in the same way you spread labor costs. Other ODC can include labor (services) and materials involved in the making and printing of drawings, photographs, negatives, photostats, printing of manuals for delivery, etc.

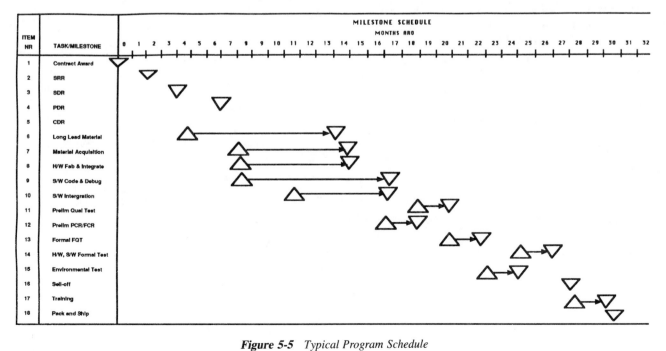

Figure 5-5 *Typical Program Schedule*

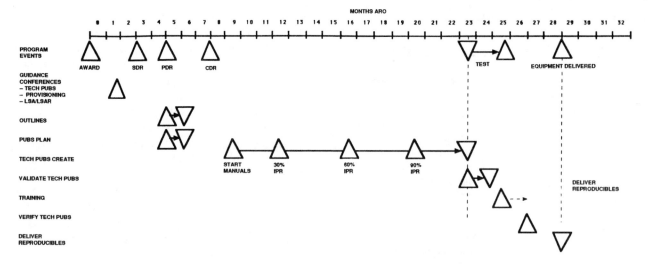

Figure 5-6 *Making a Technical Publications Milestone Chart*

Spreadsheets

Projections can be done in several ways. Typically, "spreadsheet" formats are used. You spread the number of hours bid for each labor category over the time period from start to completion of each task and then total the hours for each month. The result is a rough estimate of the number of hours per month for each labor category. Multiply 8 hours per day, × 5 days per week, × 52 weeks per year = 2080 hours. Dividing by 12 months per year results in approximately 173 hours per month.

These are but estimates, since time for vacations and holidays is not accounted for. Work for 1.3 people in reality can be handled by one person working overtime rather than having one or two people working only 0.3 hours a month. By using the chart, you can decide if there will be enough personnel to handle the tasks at peak loads and when different labor categories will be required. For instance, engineers are required to support the writing effort. Usually their time is on an as-needed basis which is spread as Level of Effort (LOE). But there are times when they may be needed full time, eight hours a day, such as during validation and/or verification of the manual. The chart also identifies when the production (printing and illustrating) personnel can expect work to start and when peak demands will hit.

Conclusion

Before writing the manual, then, the technical writer needs to re-evaluate the sources such as the proposal and contract. Based on these sources, the writer can plan a milestone schedule and make personnel projections. All of this data can be put into computers, so that it is constantly controlled and periodically updated. Writers in part assume the role of managers who are responsible for getting the tasks done on time and to the satisfaction of the customer.

Table 5-3 *Labor Categories in Technical Publications*

Category	Function
Technical Writer	Prepares material (text, tabular and art) for technical manuals.
Illustrator	Prepares pencil/ink drawings, photos, computer graphics from rough sketches and other source data (such as engineering drawings) for inclusion in the technical manuals.
Typist/Word Processor	Prepares all material in typed format that is contained in the manual.
Proofreader	Proofreads all material prepared by typist/word processor
Editor	Edits all material prepared by the writer, either before or after it goes into typing/word processing.
Engineer	Provides source material to writer; reviews material for technical accuracy and adequacy; participates in validation/verification.
Clerk	Provides general clerical support, including typing and filing
Planner	Participates in preparation of cost and schedule material and program reviews.

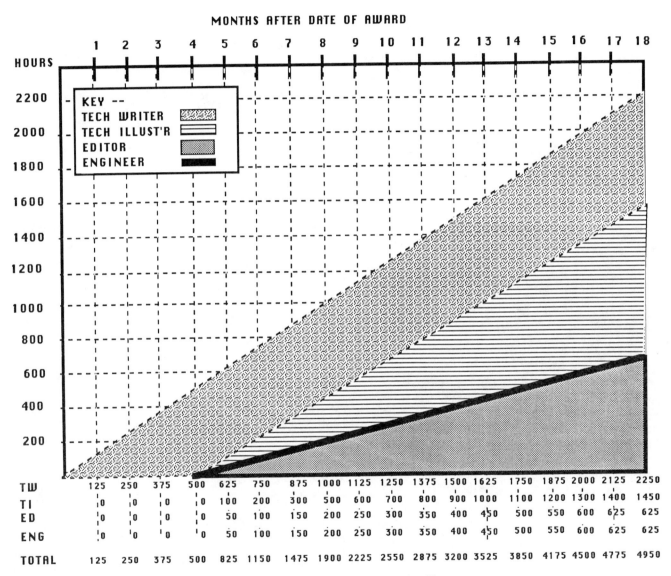

MONTHS AFTER DATE OF AWARD

	1	2	3	4	5	6	7	8	9	10	11	12	13	14	15	16	17	18
TW	125	250	375	500	625	750	875	1000	1125	1250	1375	1500	1625	1750	1875	2000	2125	2250
TI	0	0	0	0	100	200	300	500	600	700	800	900	1000	1100	1200	1300	1400	1450
ED	0	0	0	0	50	100	150	200	250	300	350	400	450	500	550	600	625	625
ENG	0	0	0	0	50	100	150	200	250	300	350	400	450	500	550	600	625	625
TOTAL	125	250	375	500	825	1150	1475	1900	2225	2550	2875	3200	3525	3850	4175	4500	4775	4950

Figure 5-7 *Typical Personnel Projection Chart*

Chapter 6

The Real Work Starts: Actually Writing the Manual

Introduction

After re-evaluating all the sources, the actual writing of the manual can then begin. Many people are involved in creating technical manuals. In addition to technical writers, there are engineers, technicians, illustrators, word processors/typists, editors/proofreaders, photographers, printers, etc. Technical manuals go through a series of steps before they reach camera-ready form ready for printing. Figure 6-1 is a flow diagram which illustrates the typical sequence by which the technical publications organization and its supporting functions prepare the data. The functions have various names, such as Art Department and Production Department, etc. The particular names for each company's functional area may vary.

Analyzing the Source Data

Source data is data the technical writer uses to prepare the technical manual. It can include the following:

Engineering drawings:
- Assembly (top, front, rear view, etc.)
- Cabling/Wiring
- Schematics
- Logic Diagrams
- Power Distribution Diagrams
- Parts Lists

Engineering data:
- Preliminary Design Review (PDR) and Critical Design Review (CDR) packages
- Engineering sketches and notes
- Data compiled from interviews with engineers
- Data from existing manuals
- Vendor Data
- Acceptance Test Procedures
- Reliability/Maintainability Data
- Logistic Support Analysis Data and Records (LSA/LSAR)
- Provisioning Data

At the start of the writing task (usually after the CDR), much of this source data is preliminary and not entirely accurate. Engineers are still designing hardware, and drawings are just starting to be formally released (put under drawing control). However, there is enough general information worth sorting to start writing.

PDR and CDR Reviews

PDR and CDR packages are usually bound copies of various technical descriptions including Prime Item Development Specifications, Critical Item Specifications, and viewgraphs the contractor prepares to give presentations of program status to the customer. These packages represent two of the initial program milestones. Since PDR occurs before CDR, the data contained in the PDR package will not be as current as that of the CDR package. You, as a writer, end up using only a small portion (if any) of the PDR data as reference material since it only represents hardware at its early stages of development. On the other hand, the CDR data can include many bits of information you will find useful, including the following:

- Front panel sketches showing controls and indicators
- Top assembly and 3/4 view diagrams showing the assembly and subassembly locations
- Rack elevation drawings (drawings depicting front views of racks and identifying the location of assemblies within the racks)
- Simplified block diagrams showing signal flow, power distribution, and cabling
- Simplified logic diagrams showing circuits of assemblies
- Brief descriptions of hardware, physical, electrical and operational characteristics
- Brief functional descriptions of chassis and assemblies
- Maintenance concept
- Test data, including availability and use of Built-In-Test (BIT) and Built-In-Test Equipment (BITE)

Not all the available data contained in engineering drawings will be used as source data to prepare the manuals. Many detailed part drawings, Specification Control Drawings, Source Control Drawings, Printed Wiring Boards, Master Pattern Drawings, etc., are used for fabrication (building and assembling the hardware). In addition, all engineering drawings are not available at the start of the program. The sequence for drafting and releasing drawings is dependent upon the program plan for fabrication and assembly of the hardware. If an item is to be built using purchased parts, the drawings and parts identifications will have to be done early in the program, especially if they are "long lead" items (ones which require a significant length of time to procure).

A technical writer needs to sort data and evaluate where

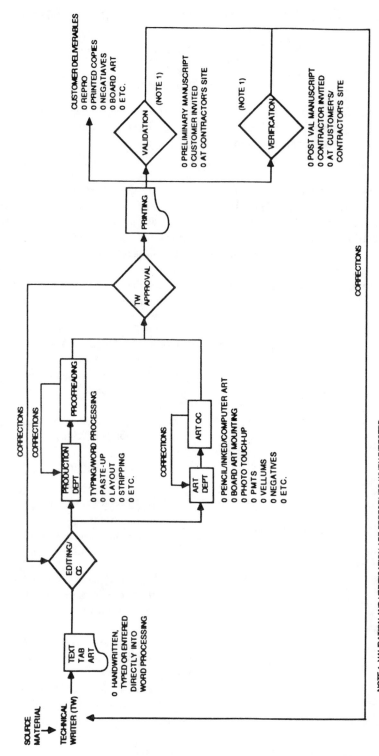

Figure 6-1 Typical Technical Publications Organization—Technical Data Flow Chart

it fits in the manual. If the contract states LSA and LSAR are to be delivered (data prepared by the Logistics Engineers), you as a writer may be required to use this material in the Maintenance Chapter. Typical LSAR data includes removal, replacement, alignment, adjustment, test and troubleshooting information about the hardware. This data is entered into a data base and made available to the writers who determine where they can start writing first.

As industrial writers adopt more modern technology and computer usage expands, the manual writing tasks are becoming more integrated through the use of electronic publishing. The logistics engineers, analysts, and writers first define the detailed maintenance tasks and identify the tools, test equipment, and supplies needed. The tasks and support equipment are part of the LSAR data base. Text is extracted from the LSAR and put into the technical publications data base, tailored with word processors and transferred to a page layout processor for automatic composition and printing. Engineering drawings generated by Computer Aided Drafting (CAD) systems can be digitally accessed by writers for modification or direct use by electronically "pasting" them in the manual. Today's final manuals may be digitally delivered on tape or disk without ever being printed on paper.

Source data includes provisioning data. Provisioning is one of the Integrated Logistics Support (ILS) tasks and is the process by which the government determines which spare parts are required to support the hardware, their costs and repairability. The government determines the quantity of spares on the basis of Mean Time Between Failure (MTBF), calculated by a maintainability analysis and total quantity of hardware items purchased that can be broken down in the estimated MTBF. Part of the provisioning task includes assignment of special codes called Source, Maintainability and Recoverability (SMR) codes. These codes are central to a technical manual and the Maintenance Allocation Charts (MAC's). All these documents must agree on what parts are repairable and what parts are not. You must research and use such information when preparing the manual.

Generating Text, Tables and Art Pages

After deciding where to begin writing, you follow the prepared outline. Most outlines identify chapters and sections. Also some include primary paragraph headings, tables and illustrations.

When you have enough information to start writing, you begin by sorting and assembling the data to prepare the text, tables, and illustrations you know will be needed. Some aspects of technical manual data are better suited to tabular rather than prose structure. Tables can be used for listing, summarizing, and referencing technical data. The typical manual will need tables to list technical characteristics of the equipment. Tables can also be used to show the following information (Tables A–F):

Table A. *Major Assemblies of XYZ Equipment*

Reference Designation	Part Number	Name	Index Number

Table B. *Technical Characteristics of XYZ Equipment*

Characteristic	Data

Table C. *Applicable Documents*

Document Number	Title

Table D. *Cabling*

Reference Designation	Connector	Mates with Connector

Table E. *Controls and Indicators*

Control/Indicator	Function

Table F. *Equipment XYZ Fault Isolation*

Step Instruction	Normal Indication	If Indication is Normal	If Indication is Abnormal

Illustrations may include a 3/4 top-down view of the equipment to show what it looks like (Figure 6-2), a top view with the cover removed to show the location of major replaceable assemblies and a front and rear view of the equipment (Figure 6-3). You may also need top level (simplified) and detailed block diagrams of the equipment to show its overall function (Figure 6-4).

The example figures are typical of the types you will need. Most source information required to prepare preliminary illustrations will be available from the CDR package and from the engineering drawings. Later you may need to modify and update these illustrations to keep them current with hardware development. If the hardware is available and your contract permits, you might use photographs in place of drawings. If you plan to use photos but the hardware is not yet available, you may temporarily have to use hand-drawn sketches or a "slip sheet." The slip sheet is a temporary substitution for an illustration and contains only a title at the bottom of the page and perhaps an "art control number" (identifies each piece of art) or negative number. It may also contain a brief statement identifying the purpose of the photo (such as "front view").

EXAMPLE

CHAPTER 1

DESCRIPTION OF EQUIPMENT

1-1.   CLASSIFICATION. The message multiplexer (MUX) Unit (figure 1-1) and this technical manual are UNCLASSIFIED. However, in use, the equipment assumes the classification of companion equipment or data being processed and/or stored. The microprocessor program stored in the equipment memory is also UNCLASSIFIED.

1-2. PURPOSE. The message MUX unit (MMU) acepts messages from four, 300-baud bit synchronous (clock and data) International Teletype Association #2 (ITA #2) Baudot Code data lines and transmits the incoming data to a Model 40 printer on a message available basis. Transfer of the data to the printer is at 2400 bauds and is controlled by the printer.

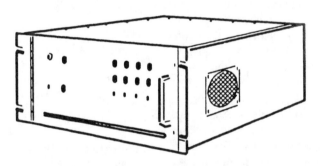

Figure 1-1. Message MUX Unit

1-3. PHYSICAL CHARACTERISTICS. The MCU is housed in a chassis that is 8.72 inches high, 17.50 inches wide and 16.50 inches deep. All controls and indicators are on a 19.00-inch wide, hinged front panel assembly. Cooling and air circulation is supplied with a fan on the side of the unit.

1-1

Figure 6-2   *Typical 3/4 Top View*

EXAMPLE

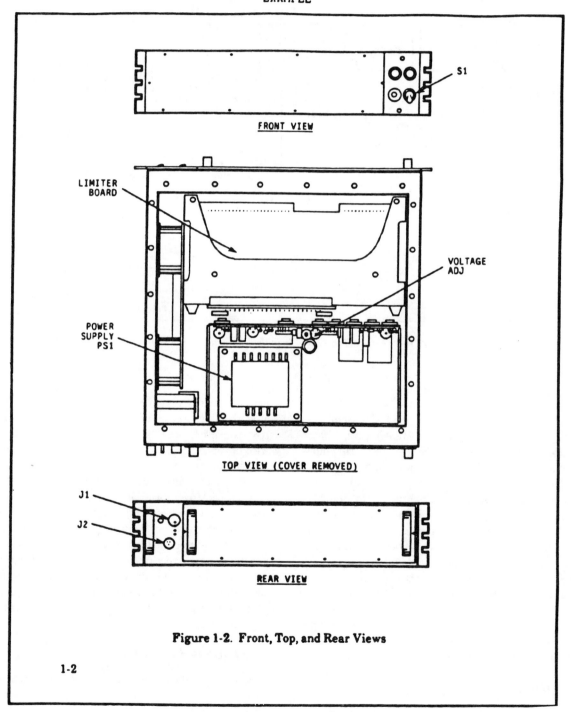

Figure 1-2. Front, Top, and Rear Views

1-2

Figure 6-3 *Typical Top View, Cover Removed*

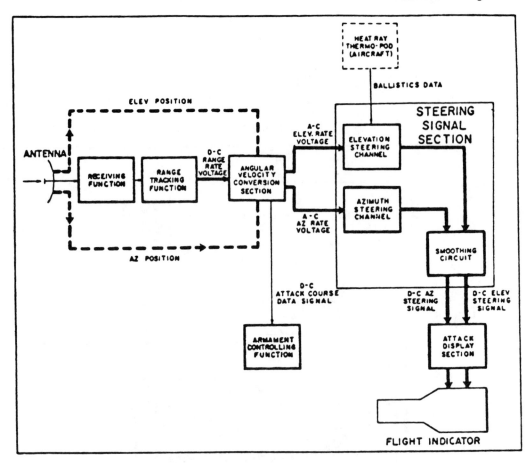

Figure 6-4 *Typical Top-Level, Simplified Functional Block Diagram*

When describing hardware, "callouts" are used to identify key areas and locations of assemblies, subassemblies, and parts of the diagrams. A callout is a descriptive block of data labeling such as "ASSEMBLY A" or a "find number." Find numbers (1,2,3) may be used in place of data blocks. The callout is connected to a particular area of the illustration by a "leader line" which points to (touches) that part of the drawing where the item is located. Everything called out in an illustration should be discussed in the text. For example, when discussing a block diagram, every block should be mentioned. Incomplete or inconsistent descriptive material may confuse the reader.

If you decide to use an illustration from another document, make sure you are aware of all copyright laws. The use of "boilerplate" material (the multiple use of the same material) is a technique to save time and to maintain consistancy in certain areas of a manual.

Freeze Date

A "freeze date" is the date when engineering design change information will no longer be accepted; design (so far as the manual is concerned) is frozen. Beyond that date, no new information is accepted. The freeze date is normally set somewhere between 30 and 90 days before the prelimi-

nary manuscript is printed. Design changes to the hardware after the freeze date can be included in the manual after the manual has been validated. By combining the changes with the corrections, you save time in writing, illustrating, and producing the manual.

Remember, every time you accept new information, you must go back over material already prepared and insert the new data. At the beginning stages of the manual preparation, adding and changing material does not take too much effort. But as more material is written, even a small change can require extensive rework. It takes more time to incorporate changes in a larger volume of text and artwork. With each additional change, your productivity in terms of hours per written pages decreases and the cost of writing each page increases.

In-Process Reviews

As you write the manual, the customer may want to check your progress. The contract may cite specific intervals or progress milestones (i.e. 30 percent, 60 percent, 90 percent as shown in Figure 5-6) for checking the manual to see if you are on schedule and if you are preparing the material according to the specifications and approved outline. The first In-Process Review (IPR) gives customers their first look at what you are writing and how you are organizing the manual. Before this time, they may have seen only the manual's outline.

Since IPR's are formal meetings, all comments are recorded by either the customer or contractor. Some customers use specific forms to record comments (i.e. the Air Force AFTO Form 158). At the end of the IPR, the key personnel attending the meetings (lead writer and customer) usually sign the meeting minutes. You as a technical writer working on the project will be expected to respond to the comments by the next scheduled IPR. Also, a specific time limit may be contractually set within which you may respond to comments.

At the next IPR, previous IPR comments are reviewed and your responses are checked. If for any reason you decided that one or more of the comments were not valid, you must give reasons to justify your position. The second IPR may result in more comments than the first because by this time you will have more material available for the customer to review.

Conclusion

When starting to write the manual, you must collect and analyze source data. As writing progresses, one or more IPRs may be held during which the customer reviews progress to date. Each review contains more data and incorporates customer comments from previous reviews. You should be aware that during the entire writing procedure, you may work with several customer representatives. As a result any new representative may not totally agree to or accept previous decisions about the format or content of the manual. Changes that have a direct impact on the cost and schedule should be passed on to upper management. Upper management, in turn, may decide to accept or not to accept the customer's changes because you as a writer may not always be in a position to make final decisions on cost and schedule. The question of authority is central to any decision. If you have been given explicit authority by your company, you may have to make some hard decisions concerning cost and schedule changes to the manual.

Chapter 7

Preparing the Preliminary Manuscript

Introduction

As the manual goes through its stages of completion, it is referred to by different names such as "preliminary," "draft preliminary," "draft," "final draft," "manuscript," "camera-ready," and "repro." Names used to describe two different manuals written to the same specifications for the same equipment at the same stage of development may be different, depending on the customer.

Definitions

Basically, preliminary, draft, and manuscript mean the same thing. The names for each stage of a manual's production are cited in the contract. A preliminary manuscript according to MIL-M-38784 is an interim form of a manual and may lack some technical details. It is fully edited and printed on one side of a page in a double-spaced, single column, and justified on the left-hand side (Figure 7-1). The reproduction, cover and binding must result in a copy suitable for use during the preliminary review, validation and perhaps any training session.

Camera-ready copy (CRC) is generally agreed to be that stage of the manual which is 100 percent complete and validated, in final format and ready for printing. It is the last step of the manual's evolution, excluding future changes and revisions. None of the other forms (draft or preliminary) of the manual are usually used for reproduction and disbursement to the users to support the intended hardware. On rare occasions, the preliminary or draft version of a manual may be shipped with hardware to the users because the manual has not yet been completed. In this instance, the manuals are stamped "Preliminary" so the user knows that the manual has not been finished and that another copy will be sent at a later date.

Text Format

The format of a preliminary manual may differ from that of a CRC. Figure 7-1 (Figure 9 of MIL-M-38784) has a relatively large amount of "white space," or space between the lines since the manuscript is double spaced. Figure 7-2 in this chapter (Figure 5 of MIL-M-38784) is an example of a double-column unjustified CRC, and Figure 7-3 (Figure 5 of MIL-M-38784) is an example of a single-column unjustified CRC.

The double space of a preliminary manuscript gives the reviewer room to make notes and corrections. One and a half inches of blank space above and below the "cutlines" are provided to make the reference stand out. The specification requires that a figure be placed as close as possible following its first reference in the text. With all text and tables double-spaced, more pages are needed than in a single-spaced manual. Approximately 50 percent more text (number of words) can be placed in a single-spaced page (Figure 7-3) than in a double-spaced page (Figure 7-1).

Manuscript pages may contain illustrations rather than cutline references. Illustrations that are in the "final" or complete stage can be prepared by several methods.

- Black-on-white photostats (stats)
- Positive Image Transfers (PMTS)
- Electronic scanning
- Electrostatic copying
- Computer generated and electronically "pasted"

The illustration is located immediately after its first reference in the text. Running heads and feet (marginal copy or information placed outside the image area such as page number, technical manual publications number, etc.) are usually included on the page as well as the figure number and title. Paragraph 3.2.2.3 of MIL-M-38784 states that marginal copy can include change and security information as well as other identifying information.

Table Format

Tables for the preliminary manuscript are basically the same as for the repro or camera-ready copy. The difference is only in the spacing between lines. As shown in Figure 7-4 in this chapter (Figure 8 of MIL-M-38784), double-spaced lead lines have a lot of white space. When the manual is prepared in its final, reproducible format, that space is no longer necessary. You need to read the specifications carefully to know the requirements for tables.

Art Format

Illustrations in the preliminary manual may be anything from hand-drawn sketches to inked art or to computer-generated art, depending on what the customer wants or will accept. By the time a manual is ready for validation, the art should be at least in pencil form prepared by using standard artist's tools (straight edge and templates, etc.) or computer generated. Again, you need to read the specifications to determine requirements for mechanical lettering and line weight.

MIL-M-38784B

CHAPTER 16

SIGHTING AND FIRE CONTROL EQUIPMENT

Section I. INTRODUCTION

16-1.   SCOPE. This chapter contains information on the arrangement of the ......

*   *   * ‾ *   *   *   *   *   *   *   *

16-2.   ORIENTATION.

16-2.1   General. The oil gear M3, because of its additional features of self-synchronous operation and the self-contained hydraulic stops for elevation, requires painstaking operation.

NOTE

The procedure outlined in b and c below should be followed carefully if faulty operation is to be avoided.

*   *   *   *   *   *   *   *   *   *   *

16-2.2   Aximuth Gear. The azimuth oil gear M3 (fig 16-4) can be oriented as follows:

a.   Level the gun carriage

b.   Open the azimuth boresighting clutch by moving the boresighting level to the UP position.

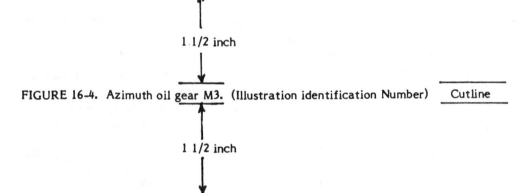

Cutline   FIGURE 16-4. Azimuth oil gear M3. (Illustration identification Number)   Cutline

c.   Depress the power synchronizing mechanism to start the oil gear.

16-1

Figure 7-1 Example of Typical Manuscript Page

The Production Process

If you as a technical writer are directly involved in printing the manual, you should be familiar with every step of the production process. When material is submitted to the Production Department, you must give them specific instructions on how you want the text prepared (i.e. the format for chapter and section divisions, headings, tables, figures and page numbering, etc.). Specifications and standards cited in the contract define format requirements which you must follow. If MIL-M-38784 is the standard cited, you must be familiar with all aspects of the requirements covered

MIL-M-38784B

4-3 PREVENTIVE MAINTENANCE PROCEDURES

The preventive maintenance procedures listed below provide information necessary to conduct a comprehensive program of cleaning and inspecting the AN/SPS-10. Each procedure includes equipment and materials required and step-by-step instructions on how to perform the preventive maintenance.

```
CAUTION
```

Comply with Navy Safety Precautions for Forces Afloat, OPNAVINST 5100 series prior to performing preventive maintenance.

4-3.1 AIR FILTER CLEANING PROCEDURES. The air filters in the Receiver-Transmitter, the Modulator, and the Video Clutter Suppressor (AN/SPS-10 Field Change No. 22, 30, or 31 incorporated) should be cleaned monthly.

4-3.1.1 Tools and Equipment Required.

1. Warning tags
2. Vacuum cleaner with non-metalic nozzle

```
WARNING
```

High voltages that are dangerous to life may be stored on capacitors after power is removed.

4-3.1.2 Preliminary Actions.

1. Turn OFF and tag radar bulkhead main power switch in radar equipment room.
2. Locate filter in center underside of Receiver-Transmitter cabinet.
3. Locate filter on right side below connector panel on Modulator cabinet.

4. For equipments with Video Clutter Suppressor MX-8756A/SPS-10, locate filter inside cabinet door.

4-3.1.3 Procedures for Cleaning Air Filters.
1. Remove filters.
2. Vacuum filters, reversing normal air flow.
3. Inspect filters for cleanliness. If additional cleaning is required:
   a. Wash filters in solution of warm water and detergent
   b. Rinse filters in clean, fresh water.
   c. Blow excess moisture from filters with low-pressure air.
   d. Allow filters to dry thoroughly.
4. Reinstall cleaned filter.
5. Return equipment to normal readiness condition.

4-3.2 AS-936( )/SPS-10B ANTENNA ASSEMBLY AND OIL LEVEL INSPECTION, AND LUBRICATION OF ANTENNA DRIVE MOTOR. These maintenance procedures should be performed quarterly, when AS-936( )/SPS-10B Antenna Assembly (Units 19, 20, or 21) is installed.

4-3.2.1 Tools and Equipment Required.

1. Clean rags
2. Warning Tags
3. Small funnel
4. Safety harness
5. Oil, MIL-L-9000 or MIL-L-17331
6. Grease, MIL-G-23827
7. Grease, MIL-G-81322
8. 8" adjustable wrench
9. 3/4" fill pipe with grease cap

4-3.2.2 Preliminary actions.

1. Comply with ship's regulations for working aloft.
2. Turn off and tag radar bulkhead main power switch.
3. Press STOP button on Manual Controller Switch and tag "MAN ALOFT".
4. Turn Antenna Switch Control to OFF RESET.

Figure 7-2 *Example of Double-Column, Unjustified Text*

MIL-M-38784B

Publication Number

CHAPTER 3

HANDLING AND STORAGE

3-1. GENERAL. Compliance with AFR 127-100 and the instructions in this manual will ensure safe handling, storage, and serviceability of widgets. Waivers and deviations will be in accordance with AFR 127-100. Stored widgets should be protected from adverse climactic conditions. The main hazards linked with the storage and handling of items listed in this TO are:

    a. Blast.

    b. Fragments.

    c. Fire.

3-2. SPECIAL TERMS. The following terms, as defined, apply to widgets.

    NOTE

Shelf and service lives are not cumulative. Any combination of shelf and service life accrued by an item cannot exceed the shelf life.

3-2.1. Shelf life: The length of time an item can remain in storage. The expiration date for shelf life on items with the month and year listed is the last day of the month.

3-2.2 Service life: The length of time an item can remain in operating configuration or in actual usage.

NOTE: For those items packed in hermetically sealed tear strip containers service life starts on date of opening and continues until item(s) are expended.

3-2.3 Magazine: Any building or structure, except an operating building, used for storage of explosives, munitions, or loaded munition components.

3-3. IDENTIFICATION. The use of standard nomenclature and lot number/serial number is mandatory for all storage records and communications. Legible identification markings will be kept on munitions in storage.

Figure 7-3 *Example of a Single-Column Page*

in Paragraph 3.2 of that spec. You must also understand the order of precedence of all specifications that are applicable to your manual. This is especially necessary when conflicts exist between the specifications.

The Production Department

A typical Production Department does word processing, paste-ups, layouts, cutting and stripping, ruling tables, and other such tasks. The writer usually turns the preliminary manuscript over to a coordinator, who may also be an editor/proofreader, or quality controller. This person distributes art to illustrators and text to word processors. When the text is typed and proofed and the graphics are completed,

the coordinator assemblies the document. Material consisting of formatted text, tables, and art is then sent back to writer.

The Production Department usually assumes the manuscript is technically accurate. The department is less concerned with the contents of the text than with the details of page formatting, such as page borders (top, bottom, left and right margins), centering, justifying left or right, and table headings. MIL-M-38784 contains several figures that identify format requirements and gives examples of pages prepared to the proper requirements. Figure 7-5 in this chapter (Figure 3 of MIL-M-38784) gives instructions on style, size, capitalization, leading, and spacing. In the same spec, Figures 1, 2, 4 through 26, 37 through 42 and 44 through

MIL-M-38784B

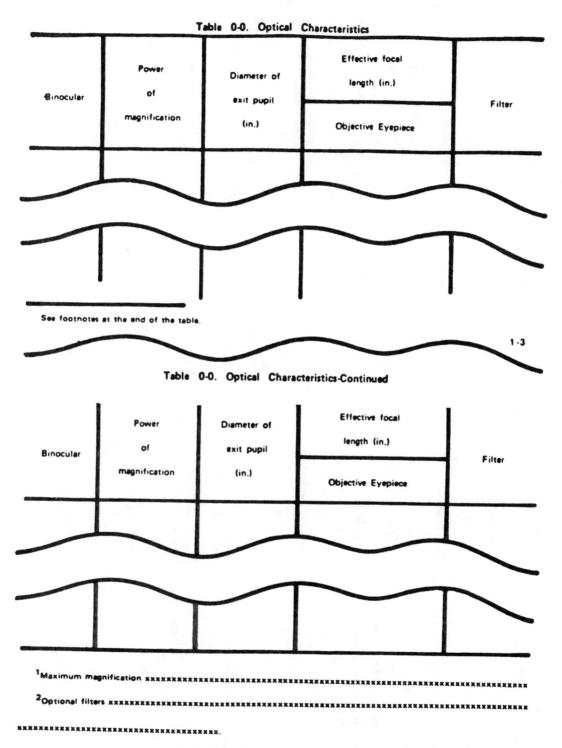

Figure 7-4 *Example of Tabular Structure*

MIL-M-38784B

USE	TYPESET	CAPITALIZATION	LEADING (POINTS)	SPACING
1. Marginal Copy	Futura Demibold 10 Futura Bold Italic 12 Gothic or Roman Bold or Extra Bold	Upper and Lower Case	2	6-Point Above or Below Text
1a. Publication No. or Page No.	Same as Above	Upper Case		6-Point Above or Below Text
1b. Change Number	Futura Demibold 10 Gothic Bold 10	Upper and Lower Case		6-Point Below Text
1c. Security Class.	Futura Bold 14 *	Upper Case		6-Point Above or Below Text
1d. Page Content/ Equipment Identification	Futura Demibold 10	Upper and Lower Case	2	6-Point Above Text
1e. Deleted Page(s) Notation	Futura Demibold 8	Upper and Lower Case	2	6-Point Above or Below Text
2. Text	Bodoni Book 10 Garamond Bold 10 Century Textbook 10	Upper and Lower Case	1	12-Point Above or Below Illustration or Table 6-Point Above Warning, Caution or Note Heading 6-Point Below Warning, Caution or Note Heading
2a. Emphasis	Futura Bold Italic 10	Upper and Lower Case		
2b. Formulas and Equations	Century Textbook Italic 10 Cheltenham Old Style Italic 10	Upper and Lower Case	1	12-Point Above or Below Text, Illustration or Table
3. Part Nos., Chapter Nos. and Titles; Appendix and Alphabetical Index Headings	Futura Demibold 14 * Futura Bold Condensed 14 *	Upper Case	6	48-Point Below Marginal Copy 18-Point Above Text, Illustration or Table
4. Section Number and Title	Futura Demibold 12 Futura Bold or Extra Bold 14 *	Upper Case	6	24-Point Below Chapter Title, when Applicable 28-Point Below Marginal Copy. when Applicable 18-Point Above Text, Illustration or Table
5. Paragraph Headings 5a. Primary Sideheads	Futura Demibold 12 Futura Bold or Extra Bold Condensed	Upper Case	2	8-Point Below Preceding Text, Part, Chapter, Section Title or Marginal Copy
5b. First Subordinate Sidehead	Futura Bold Italic 10 Futura Bold or Extra Bold Condensed 10 Roman Bold 10	Upper and Lower Case	1	4-Point Below Preceding Text or Marginal Copy

Figure 7-5 *Example of Typeset Text*

	USE	TYPESET	CAPITALIZATION	LEADING (POINTS)	SPACING
5c.	Second Subordinate Sidehead	Century Textbook Italic 10 Roman Bold 10	Upper and Lower Case	1	4-Point Below Preceding Text or Marginal Copy
5d.	Third Subordinate Sidehead	Century Textbook Italic 10 Roman Bold 10	Upper Case for First Letter of Each Word	1	4-Point Below Preceding Text or Marginal Copy
5e.	Fourth Subordinate Sidehead	Century Textbook Italic 10 Roman Bold 10	Upper Case for First Letter of Each Principal Word	1	4-Point Below Preceding Text or Marginal Copy
6.	Figure No. and Title	Century Textbook Italic 8 Futura Demibold Italic 10	Upper Case for First Letter of Each Principal Word	2	6-Point Below Illustration
7. 7a. 7b.	Legend (Key) On Artwork In Text	Century Textbook 8, Bodoni Book 8 Same as Above	Upper Case Upper Case for First Letter of First Word	 1 1	Set Solid Above Caption 6-Point Below Illustration 6-Point Above Figure Title Set Solid Above Legend Line
8.	Warnings (Heading)	Century Bold Italic 10 Futura Extra Bold (Boxed)	Upper Case		4-Point Above and Below Text
9.	Cautions (Heading)	Same as Above	Upper Case		4-Point Above and Below Text
10.	Notes (Heading)	Cheltenham Old Style Italic 8 or 10 Futura Extra Bold 10	Upper Case		4-Point Above and Below Text
11.	Footnotes	Century Textbook 8	Upper and Lower Case	1	2-Point Below Table. Full measure at Bottom of Page Separated by a 1-inch hairline
12.	Table No. and Title	Century Textbook Italic 8 Futura Bold Italic 10	Upper Case for First Letter of Each Principal Word	2	Center Above Table. Repeat on Continued Pages
13.	Table Column Heads	Century Textbook 10 Bodoni Book 8, Futura Bold 10	Upper Case or Upper and Lower Case	1	
14.	Table Text	Century Textbook 10 or 12 Futura Bold or Medium 8	Upper and Lower Case	2	
15.	Boxheads	Century Textbook 8	Upper Case or Upper and Lower Case		Repeat on Continued Pages
16.	Rules	Hairline 2			
17.	Parts Lists Column Heads	Century Textbook 8	Upper Case	1	
18.	Parts Listings	Century Textbook 8 or 10 Futura Bold or Medium 8 or 10	Upper and Lower Case	11	

*If 14-point is not available, the next smaller size is permitted.
All type sizes may be plus or minus one point.
Slight variations in leading and spacing are permitted.
Reproducible copy, that will require no reduction, shall use above type sizes.
Reproducible copy prepared oversize shall use type that will reduce to specified sizes.

IT IS NOT THE INTENT OF THIS SPECIFICATION TO SPECIFY, THE METHODS OR COMPOSING EQUIPMENT TO BE USED, BUT ONLY TO SPECIFY REQUIRED RESULTS.

Figure 7-5 *Continued*

46 provide further examples of text, tables, and front matter. Today, the pages of a manual can easily be formatted through the aid of desk top publishing. Writers who are knowledgeable of their computers and word processing applications can write and format an entire manual while sitting at their desks.

The Art Department

The Art Department receives illustrations in various forms from the coordinator. These illustrations may be hand-drawn sketches, marked-up engineering assembly drawings, marked-up photographs, copies of pages from other documents, such as the Preliminary Design Review (PDR) or Critical Design Review (CDR) packages, and computer graphic art. Typically, the Art Department prepares computer-generated and pencil and inked art, does board-mounting, photo touch-up, layouts and many other art functions, including handling of reproduction of art in the form of photostats, vellums, etc.

The Art Department passes the original art (not computer generated) to the photo lab or print center. The reproductions are returned to the Art Department, which then performs a quality-control (QC) check. After all art is prepared, it is returned to the Editor/QC person, who returns the document and art work to the technical writer. The writer then checks the document for format and content. Any errors are marked and returned for rework. If everything is acceptable, the material will be submitted to the company's printing center along with special instructions on how to print (type size and weight of paper), the number of copies wanted, type of covers, and method of binding.

Editing

MIL-HDBK-63038-2 (TM), Section 19 defines editing as follows:

Editing: Checking manuscript for technical accuracy, conformity to a particular standard of data presentation, and clerical accuracy. Editing is also required to bring a manuscript into the form required for production to obtain a satisfactory final product.

Editing includes doing whatever is necessary to make the manuscript a good quality, usable product conforming with the specification, contractural and style-guide standards. The editor is responsible for checking all inaccuracies, all violations of the style guide, all violations of specifications or applicable deviations to the specification.

According to this definition, editing means revising the manual until it is technically accurate, until it complies with set standards, and until it exhibits grammatical accuracy. Editing is an on-going process, from the preliminary manuscript stage through the intermediate review stages to the final camera-ready/repro copy. Your goal as a technical writer is to produce a manual that is consistent in style and language, point of view and format, while at the same time making sure the information in the manual is both accurate and readily accessible to the intended reader. Most importantly, the manual must comply with the format and content standards and specifications required by the governing contract.

Conclusion

Creating the preliminary manuscript is a repetitive cycle of first preparing the text and art, proofreading it, resubmitting it for corrections, recorrecting, reproofing, and again resubmitting it for final printing. Keeping down the number of times art and text must be corrected depends on the experience and skill of the personnel in the Production Department and the accuracy of the manuscript. If problems arise in getting the manual prepared in the time allotted, you may have to inform management so the problems can be corrected as soon as possible.

Chapter 8

Validation, Verification, and the Completed Manual

Introduction

Manuals undergo rigorous review activities before they are signed off as completed. The two main review activities are called validation (a contractor in-house review) and verification (a customer in-house or out-of-house review). This chapter explains these two activities.

Validation

Validation is that "hands-on" activity which the contractor does to confirm the accuracy and adequacy of data in the technical manual. A Validation Record, which may be part of a Contract Data Requirements List (CDRL) deliverable, is used to record changes needed to correct errors and omissions in text and art that are uncovered during the validation. DI-M 33408/H-108-1 cites the following instructions for validation.

1. A Validation Completion Report shall be completed at conclusion of each validation effort for operating and maintenance procedures that are tested by actual performance or simulation in accordance with the terms of the contract. The report will be prepared on AFSC Form 11, Validation Completion Report, and unless otherwise directed, will be available at the time the manual is delivered to the government.

2. A technical order is properly validated when the following conditions are met:
   a. Contractor's engineering review has been completed.
   b. Technical manual procedures can be used to operate and maintain the system/equipment as stated.
   c. Information reflects as-built configuration of systems/equipment and includes all engineering changes.
   d. Procedural instructions are understandable and adequate to perform all operations and maintenance functions.
   e. Sequences of operation and maintenance instructions are compatible with performance.
   f. Data is compatible with Qualitative and Quantitative Personnel Requirements Information (QQPRI).

3. Forms required to implement this Data Item Description may be obtained through the procuring or administrative contracting officer.

Some contracts do not require use of specific forms on which to record changes. In these cases the customers may use copies of the manuals to "red-line" (mark up) the needed changes.

Validation Schedule

Paragraphs 4.5.1 and 4.5.1.7 of MIL-M-63036 define requirements for a Validation Schedule. This schedule identifies where (the place), when (the start and stop dates), and how (the method) the validation will take place. It is submitted to the customer who reviews and either accepts it as is or makes recommendations for changes.

Validation Performance

Validation is the contractor's activity. Customers are invited to attend, but their attendance is not mandatory unless the contract or any binding documents stipulate they must attend. The purpose of the validation is to confirm the accuracy of all procedural data in the manual. Paragraph 4.5.1c of MIL-M-63036 cites the two main criteria the Army considers as an acceptable validation: (1) conformance to applicable requirements and (2) technical adequacy and accuracy. Under item (2), five key points are listed:

   (1) Agreement of maintenance data with the Maintenance Allocation Chart (MAC)
   (2) Agreement between the MAC and Parts Manuals
   (3) Illustration, essentiality and adequacy of existence
   (4) Adequacy of references to text
   (5) Instructions for manufacture, assembly, replacement and repair as directed by Source, Maintenance and Recoverability (SMR) codes

Paragraph 12.2 of the same spec restates that validation is making sure all the data agrees with each other.

If the customer elects not to be present at the validation, the impact can be both positive and negative. On one hand, contractors can proceed with the validation at their own pace, can stop to rewrite as necessary or can reschedule the whole validation or any part of it as needed. On the other hand, the contractors' engineers, technicians and technical writers who are familiar with the hardware and the manual may overlook the inexperience of the intended user. Engineers may assume the user knows how to hook up a piece of test equipment or how to make an adjustment to the equipment. As a result, when customers receive the validated manual with these built-in assumptions, their personnel might have difficulty performing their own verification. Thus the chances increase that the contractor will receive more negative comments about the manual. If the customer had been present during validation, these problems could have been detected and appropriate corrections made.

The contract may require a preliminary manuscript be delivered to the customer prior to the actual validation. The time between receiving the manuscript and starting the validation helps the customer become familiar with the content of the manual ahead of time. Customer representatives attending validation are usually restricted to technical publications personnel. The customer's illustrating and production/editing personnel do not attend the validation. When the validation is completed, the customer may want an additional period of time for non-attendees such as their editors, graphic illustrators, and engineers to review the manual's material. Formal comments are mailed to the contractor 30 or 60 days later. The contractor must respond to all the review comments either by incorporating them into the manual or by explaining why they were not incorporated.

Figure 8-1 shows a typical validation schedule. During validation, the customer may prefer to use the test equipment listed in the manual (Figure 8-1, Block 7) and cited in the Logistics Support Analysis (LSA). However, if test equipment listed by LSA was selected from the customer's present inventory of test equipment, the Preferred Items List (PIL) or Test, Measurement and Diagnostic Equipment (TMDE) Register, the contractor may not own these items. The contractor may request permission to use alternative equipment to perform the tests. If the customer still requires the contractor to use the listed test equipment, the customer may need to provide the test equipment as Government Furnished Equipment (GFE). The customer cannot normally expect the contractor to buy or rent costly test equipment for validation unless the requirement is part of the contract.

Verification

Verification is a customer activity. The sequence of events follows those of validation. The customer performs all activities and the contractor is usually required to be present in an advisory capacity. The customer may want a combined validation and verification. In this case, one review is considered to be sufficient for testing the manual's validity. All tests, adjustments, assembly, disassembly, etc. are performed by customer personnel. Comments are recorded in the same manner as during validation. The contract may require corrections to be made ''on the spot'' at the customer's production facilities while the verification is on-going. Normally, the text, tables, and illustrations are ''red-lined'' or marked up with comments and corrections and then edited into the manual after completion of the verification.

During verification, all comments recorded during validation are reviewed. You must have a record of all validation comments and responses to those comments available for the customer. The customer will analyze the responses. If

they are justified, the customer will accept them. If the customer disagrees with some responses, you will be required to make additional changes to the manual.

The Final Manuscript

After reviewing all validation and verification corrections, you need to incorporate them into the manual and create the camera-ready copy. You should be especially careful if the final repro of the camera-ready manual must be in a different format from the preliminary or draft manual. If the customer finds that data has either been omitted or formatted contrary to the cited contract, it could be quite costly to make changes after preparing camera-ready copy (CRC) and printing the manual.

When you make mistakes in the CRC, you may be required to redo it, to make new photostats of the art, to make new page negatives, and possibly to reprint a large number of copies of the manual. As a result, you will add to the cost by having to pay for more:

- Writer time to go back through the manual and incorporate the missing data
- Illustrator, editor and word processor time to rework the repro
- Proofreading time to check the retyped material
- Other Direct Costs (ODC) to pay for additional stats, negatives, printing and bindery fees
- Production time to mount stats, rule tables, and add such items as rubdowns, symbols, etc., or generate computer art.

If you fail to incorporate all the customer's corrections into the manual, the customer may reject the entire manual. Rejection means you have failed to perform, and this rejection is reported to upper management. Management, in turn, will (1) try to appease the customer and (2) turn to you for an explanation. If the customer represents a high volume of business to the company, your management may make great efforts and even sacrifices to see that the problem is resolved to the customer's satisfaction (the customer is always right).

As a corrective measure, it may be necessary for management to replace the lead technical writer who headed up the writing project. This is rather a drastic step and is not usually done unless the customer has a series of complaints against the writing task or the lead writer. A manual that has been through the In-Process Review (IPR), validation, and verification should never face total rejection.

Deliverable Materials

The contract can stipulate several types of deliverables such as the camera-ready repro, printed copies of the manual, boardmounted art, negatives, and computer documentation in some form of electronic media (tape, disk, etc.). The customer uses negatives to make reprints of the manual. The camera-ready copy is stored for future use to allow making corrections or changes to the manuals. These

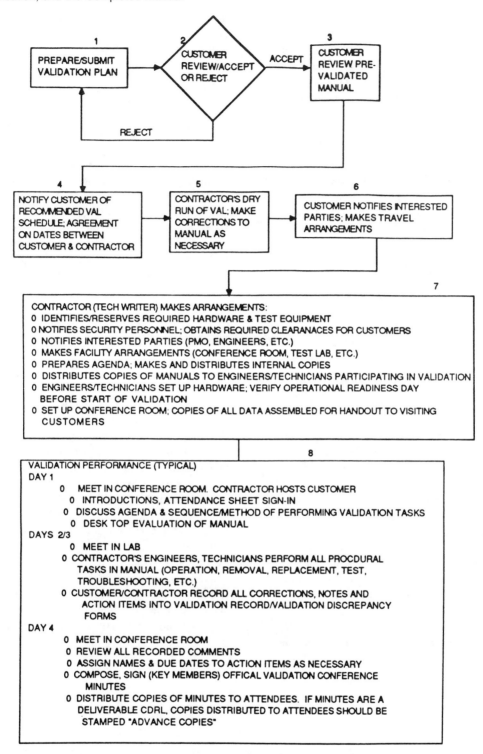

1. PREPARE/SUBMIT VALIDATION PLAN

2. CUSTOMER REVIEW/ACCEPT OR REJECT

REJECT

ACCEPT

3. CUSTOMER REVIEW PRE-VALIDATED MANUAL

4. NOTIFY CUSTOMER OF RECOMMENDED VAL SCHEDULE; AGREEMENT ON DATES BETWEEN CUSTOMER & CONTRACTOR

5. CONTRACTOR'S DRY RUN OF VAL; MAKE CORRECTIONS TO MANUAL AS NECESSARY

6. CUSTOMER NOTIFIES INTERESTED PARTIES; MAKES TRAVEL ARRANGEMENTS

7. CONTRACTOR (TECH WRITER) MAKES ARRANGEMENTS:
- 0 IDENTIFIES/RESERVES REQUIRED HARDWARE & TEST EQUIPMENT
- 0 NOTIFIES SECURITY PERSONNEL; OBTAINS REQUIRED CLEARANACES FOR CUSTOMERS
- 0 NOTIFIES INTERESTED PARTIES (PMO, ENGINEERS, ETC.)
- 0 MAKES FACILITY ARRANGEMENTS (CONFERENCE ROOM, TEST LAB, ETC.)
- 0 PREPARES AGENDA; MAKES AND DISTRIBUTES INTERNAL COPIES
- 0 DISTRIBUTES COPIES OF MANUALS TO ENGINEERS/TECHNICIANS PARTICIPATING IN VALIDATION
- 0 ENGINEERS/TECHNICIANS SET UP HARDWARE; VERIFY OPERATIONAL READINESS DAY BEFORE START OF VALIDATION
- 0 SET UP CONFERENCE ROOM; COPIES OF ALL DATA ASSEMBLED FOR HANDOUT TO VISITING CUSTOMERS

8. VALIDATION PERFORMANCE (TYPICAL)
DAY 1
- 0 MEET IN CONFERENCE ROOM. CONTRACTOR HOSTS CUSTOMER
- 0 INTRODUCTIONS, ATTENDANCE SHEET SIGN-IN
- 0 DISCUSS AGENDA & SEQUENCE/METHOD OF PERFORMING VALIDATION TASKS
- 0 DESK TOP EVALUATION OF MANUAL
DAYS 2/3
- 0 MEET IN LAB
- 0 CONTRACTOR'S ENGINEERS, TECHNICIANS PERFORM ALL PROCDURAL TASKS IN MANUAL (OPERATION, REMOVAL, REPLACEMENT, TEST, TROUBLESHOOTING, ETC.)
- 0 CUSTOMER/CONTRACTOR RECORD ALL CORRECTIONS, NOTES AND ACTION ITEMS INTO VALIDATION RECORD/VALIDATION DISCREPANCY FORMS
DAY 4
- 0 MEET IN CONFERENCE ROOM
- 0 REVIEW ALL RECORDED COMMENTS
- 0 ASSIGN NAMES & DUE DATES TO ACTION ITEMS AS NECESSARY
- 0 COMPOSE, SIGN (KEY MEMBERS) OFFICAL VALIDATION CONFERENCE MINUTES
- 0 DISTRIBUTE COPIES OF MINUTES TO ATTENDEES. IF MINUTES ARE A DELIVERABLE CDRL, COPIES DISTRIBUTED TO ATTENDEES SHOULD BE STAMPED "ADVANCE COPIES"

Figure 8-1   *Validation Schedule*

changes can be made by the customers themselves or by other companies. Boardmounted art is required with the repro in case illustration changes are needed later.

Computer-Aided Acquisition and Logistics Support

When deliverable data must meet Computer-Aided Acquisition and Logistics Support (CALS) requirements, an entirely new approach to the development of deliverable data must be undertaken. New specifications and standards are being developed which will define how technical manuals will be formatted as part of an integrated deliverable database. CALS will impact all areas of Integrated Logistics Support (ILS). Technical writers will then be able to access the data bases of both Logistic Engineering (LSA/LSAR) and engineering drawings. One of the goals of CALS is to enable every writer to access, copy, modify as necessary, and produce maintenance data and drawings (assembly, functional, wiring, etc.) in less time and at lower cost than it takes using current methods. Chapter 19 provides a brief introduction of CALS and its impact on the development of technical data.

Inspection

The customer usually inspects the CRC before you shoot negatives and before you print the deliverable manual copies. As part of this inspection the repro may also be given a final check to make sure it meets the format required in the specifications.

Printing and Binding

When the customer gives the go-ahead to print the manual, you must review the contract to check for method of binding, quality of deliverables, types of binders, etc.

The repro is formatted to the specification requirements. If the manual is standard, the usual sizes are 8-1/2 × 10-3/4 or 8-1/2 × 11 inches. The Government also has requirements for non-standard sizes for special manuals such as Preventive Maintenance Work Cards which are pocket-size maintenance manuals. Paragraph 3.2.2.2 of MIL-M-38784 explains the standard manual sizes used by the government. MIL-P-38790, *General Requirements for Printing Production of Technical Manuals*, contains information on both negative and printing requirements, including paper-stock for pages and covers, binders, dividers, and tables, inks, colors, press work and binder work, foldouts and fold-out-foldup pages. The specification includes a number of illustrations to help you understand the requirements.

Binding posts and three-ring binders are the typical methods used for binding a manual. For three-ring binders, the front cover and spine may require jackets which hold the manual identification. The contract should specify the binding requirements you must follow.

Your customer may require that you fill out and send Reproduction Assembly Sheets. These sheets contain detailed instructions for the printer. They make certain the printer will properly assemble the pages and illustrations of the manual and will fold the pages as required. The repro assembly sheets will also give you instructions if colors are used in the manual. MIL-P-38790 provides instructions and examples for filling out these sheets.

Packing for Shipment

After all deliverable materials have been prepared, they are shipped to the customer. The CDRL's give you the addresses of those who are to receive the material. Paragraph 5 of MIL-M-38784 gives the basic instructions for packing the manuscript and CRC. Also, Paragraph 5 of MIL-P-38790 gives the instructions on packaging negatives and printed copies of the manual. You need to follow these specifications to assure safe shipment of the materials to the customer. Special handling is required when shipping classified materials. You can get specific instructions for stamping and wrapping the manual from your Security Department and from the specification DOD 5520.22M.

Conclusion

What has been covered in this chapter has been the more traditional approach for preparing and printing a manual. But things are rapidly changing. Companies today are moving from the paper and pencil process to computerized preparation, layout, illustration, and composition. Using word processors, page layout software, computer-aided illustrations, and digitized photos allow the technical writer to concentrate more on the content of the manual. Revisions needed after engineering changes, validation, and verification are incorporated into the manual more easily. Customers may stipulate in the contracts that manuals be delivered on magnetic media or computer disks rather than just hard copies. In the future, interactive training and processes recorded on video for instant recall and use will become more common than the hard copy manuals produced today.

Chapter 9

Making Changes and Revisions to the Manual

Introduction

Changes come about for either technical or non-technical reasons. The customer uses documents such as the Engineering Change Proposal (ECP), found in a standard Request for Proposal (RFP) or Request for Quote (RFQ), to identify the requirements for hardware and data changes. The contractors can initiate changes to the manual because of hardware design changes. They release Drawing Change Notifications (DCN's), Engineering Change Notices (ECN's), Engineering Change Orders (ECO's), or some other official contractor document. The technical writer uses the latest released version of these documents as source data to update the manual. Non-technical changes are those which do not relate to physical hardware changes, and can be initiated when the customer decides to re-format the text, or re-identify the name of a piece of hardware referenced in the text, or restructure the manual.

Definitions

Change

A **change** is a modification to a technical manual that affects up to approximately 60 percent of the data. The customer usually determines when a change is needed to a manual. Changes are indicated by change bars or other symbols placed next to that part of the page where the change has been made. The change pages are printed and sent to the user, who removes the original pages in the manual and replaces them with the change pages. When a change is more than 60 percent of the pages, it may be more expedient to issue a revision of the manual.

Revision

A **revision** is a modification to a technical manual that affects approximately 61 percent or more of the data. The customer may require a revision for a manual even if less than 60 percent of the data needs changing; the breakpoint between change and revision is at the customer's discretion. A revision requires a 100 percent reprint of the manual (both changed and unchanged pages). Unlike a change, no change bars or other symbols are used to indicate where text, tables, or art have been modified. All previously generated change information (bars or symbols) in the original manual are removed and the original manual can be discarded.

Difference Data Sheets

Difference Data Sheets, described in paragraph 3.7.2.3 of MIL-M-38784, are used when two or more models of a basic hardware configuration (design) are manufactured. These sheets may be part of the original manual or added in response to equipment changes. The main part of the hardware manual contains data on the basic configuration. Difference Data Sheets cover variations of the basic configuration for the different models. These sheets are used only when differences in a model are minor. Difference Data Sheets would not be used when differences between models are so great that a large number of pages are required to document them.

Ways to Indicate Changes

Paragraph 3.7 of MIL-M-38784 describes the government's requirement for changes to both the manuscript and repro copies. Figures 39 through 44 of the spec are examples of how to make changes to the data. As a technical writer, you should be familiar with the requirements for numbering added material, identifying changes, and inserting data into the manual and deleting data from the manual. The use of change bars, shading or screening and pointing hands are illustrated in Figure 9-1 and Figure 9-2 in this chapter along with samples of additions and deletions to a paragraph and a figure.

The user must be able to recognize the changes in a manual. MIL-M-38784 stipulates that change bars be placed to the left or right margin of the page. Pointing hands and shading (or screening) of the entire changed area are often used on illustrations. Change pages have the word "Change" and the number of the change placed in the lower margin. Each time the page is changed, the change is re-numbered (Change 1, Change 2, etc.). A List of Effective Pages in the front of the manual will list all changes made and the effected pages. A page already containing a previously made change will have the old change indicators (change bars, pointing hands, etc.) removed, new ones placed where the new data is inserted and the new change number placed in the lower margin. In addition, Figure 9-3 in this chapter (Figure 44 of MIl-M-38784) gives an example of a Change Sheet that is required to help the user insert the change pages into the manual. The Change Sheet lists the old pages that are to be removed and which new change pages are to be inserted.

MIL-M-38784B

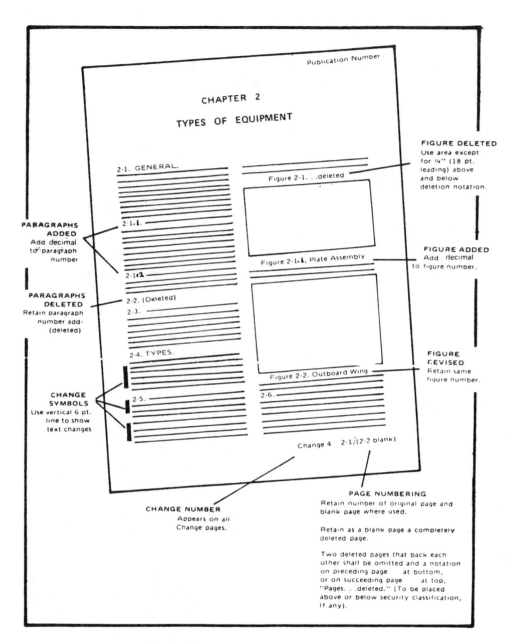

Figure 9-1 *Requirements for Change Pages*

Change Process

Change data of any significance may undergo In-Process-Reviews (IPR's) and validation/verification. When changes impart only a few pages the customer may not require interim reviews. The customer always reviews the final changes and the contractor corrects them. Deliverable data for changes include both the preliminary and final copies of the manual.

The customer may request changes be indicated in any of several forms. Changes are usually prepared in the same format as the main document. You may even need to provide informal changes to the manual marked up by hand. The insert page, paragraph, table, and figure numbers may be

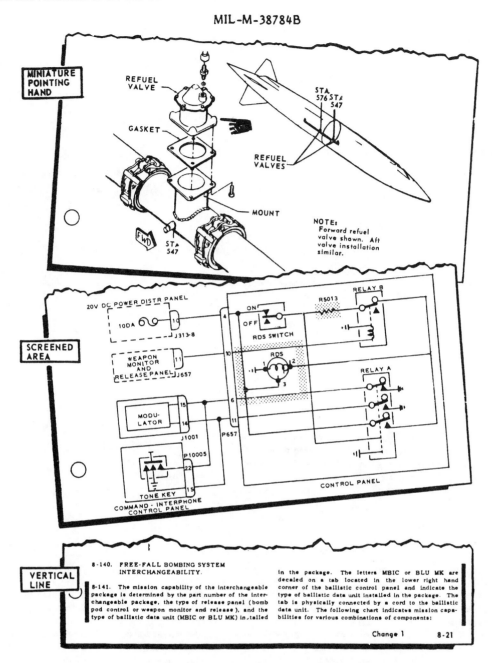

Figure 9-2 *Example of Change Symbols*

suffixed by a decimal to avoid the need to re-number all the data following the inserted material.

Revision Process

As stated previously, a revision is a re-issue of an entire manual. It contains no change bars or other change symbols. Paragraph 6.3.23 of MIL-M-38784 distinguishes between an update revision and a complete revision. The contract defines the type of revision to be prepared. Paragraph 3.8 of the same spec gives the requirements you are responsible

for in updating the manual, including re-numbering pages, paragraphs, and figures where necessary and referencing them in the text. In a revision, unlike in a change, you may be preparing the manual to an entirely new format.

Conclusion

Changes are a fact of life. Equipment changes and improves, and the documents must reflect those changes to provide useful service. The customer usually decides if an update is to be handled as a change or a revision. If the

MIL-M-38784B

TM9-4931-334-14/2
C1

CHANGE   HEADQUARTERS
                                            DEPARTMENT OF THE ARMY
NO. 1   WASHINGTON, D. C. 6 June 1972

Operator's Organizational, Direct Support

and General Support Maintenance Manual

TEST SET

RADAR AN/TPM-22

(4931-707-1229)

TM 9-431-334-14/2,5 June 1970, is changed as follows:

1. Remove old pages and insert new pages as indicated below.

2. New or changed material is indicated by a vertical bar in the margin of the page.

3. Added or revised illustrations are indicated by a vertical bar adjacent to the illustration
   identification number.

Remove Pages	Insert Pages	Remove Pages	Insert Pages
5-13 through 5-16	5-13 through 5-16	6-113 and 6-114	6-113 and 6-114
6-1 and 6-2	6-1 and 6-2	6-151 and 6-152	6-151 and 6-152
6-23 and 6-24	6-23 and 6-24	6-167 and 6-168	6-167 and 6-168
6-25 and 6-26	6-25, 6-26 and 6-26.1	6-177 and 6-178	6-177 and 6-178
6-27 through 6-38	6-27 through 6-38	6-183 through 6-192	6-183 through 6-192
6-43 through 6-46	6-43 through 6-44.3 through 6-46	6-195 through 6-198	6-195 through 6-198,
6-51 through 6-56	6-51 through 6-56	6-199 and 6-200	6-199 and 6-200
6-75 through 6-80	6-75 through 6-80, 680.1 and 6-80.2	6-213 through 6-216	6-213 through 6-216
6-81 through 6-88	6-81 through 6-88	6-219 through 6-224	6-219 through 6-224
6-95 and 6-96	6-95 and 6-96	B3 and B4	B3 and B4

File this change sheet in front of the publication for reference purposes.

Figure 9-3 Sample Change Sheet

extent of changes to a manual is less than 60 percent, the update is usually handled as a change. The contractor then provides change pages on which new information is marked in accordance with the applied specification. If the extent of changes is more than 60 percent, the entire manual is usually revised. Revisions have no change indicators except the revision level. All indicators (change bars, shadings, etc.) of previous changes are removed. In each case, changes or revisions are developed in accordance with specific requirements established by the contract.

Part II

TYPICAL CHAPTERS AND DRAWINGS
IN THE MANUAL

Chapter 10

The Description Chapter in the Manual

Introduction

The first chapter of an equipment manual describes both the manual's content and the equipment. This chapter combines photographs, line drawings, and clear, concise descriptive text to introduce the system to the users. The contents include the system's purpose and basic principles of operation, tables and/or charts that list major parts, key operational and performance characteristics, accessories, items needed but not supplied with the system and all related documents. An overview of the entire manual is also presented with this general system description.

Sections in the Chapter

Specification guidelines may vary with each writing task for each customer. Many of the typical specifications were discussed in Chapter 3. Some of them specify the content and format of the introduction or general description chapter. The material covered in these specifications differs mostly in form and order. For example, MIL-M-63036 (TM) establishes the following order for the Description or Introduction Chapter:

Section I   General Information
Section II   Equipment Description
Section III   Technical Principles of Operation

Section I tells the user about the manual, the equipment nomenclature, and the abbreviations and special terms used. Section II describes the equipment, shows the user where the major elements are, and identifies differences and key equipment data. Section III gives the user a top-level functional description of the equipment.

Section I   General Information

The General Information section provides standard and specific information describing the manual to the reader. Standard information is common to most manuals and includes forms, records, first-aid steps, and security classification information. Some information is specific to each system and includes warranty terms and alert messages (warnings, cautions, notes). The general information in Section I includes the following:

1. Scope
2. Maintenance Forms and Records
3. Hand Receipt Manuals

4. Reporting Equipment Improvement Recommendations (EIR)
5. Warranty Information
6. Nomenclature Cross Reference List
7. List of Abbreviations
8. Glossary
9. Security Classification Marking

1. Scope

The scope paragraph(s) tells the user what the manual is about. In general the scope explains the manual type and the model number(s) of the equipment covered by the manual and describes what the equipment does: its usual purpose and special limitations. It points out any special instructions added to the manual and any required elements that are omitted because they are not applicable.

2. Maintenance Forms and Records

MIL-M-63036 (TM) firmly specifies the contents of three paragraphs and defines the content and format of each of these paragraphs. The paragraphs cover the (1) Maintenance Forms and Records, (2) Hand Receipt Manuals, and (3) Reporting Equipment Improvement Recommendations (EIR). This type of information is referred to as ''Government Furnished Information'' (GFI) and is also called ''boilerplate'' information because every manual written to this specification must contain the same words. The first specified content paragraph is Maintenance Forms and Records:

- Maintenance Forms and Records
  Department of the Army forms and procedures used for equipment maintenance will be those prescribed by TM 38-750, The Army Maintenance Management System (TAMMS).

If specified by the procuring activity (the customer), the manual may also include samples of the forms most often used by the user.

3. Hand Receipt Manuals

The second specified content paragraph is:

- Hand Receipt (-HR) Manuals
  This manual has a companion document with a TM number followed by ''-HR'' (which stands for Hand Receipt). The TM X-XXXX-XX-HR (add applicable

TM number) consists of preprinted hand receipts (DA Form 2062) that list end item related equipment (i.e. COEI, BII, and AAL) you must account for. As an aid to property accountability, additional-HR manuals may be requisitioned from the source in accordance with procedures in Chapter 3, AR 310-2:

The US Army Adjutant General Publications Center
ATTN: AGLD-OD
1655 Woodson Road
St. Louis, MO 63114

4. Reporting EIR

The third specified content paragraph is an example of Reporting EIR:

- Reporting Equipment Improvement Recommendations (EIR)

  If your (insert equipment short item name) needs improvement, let us know. Send us an EIR. You, the user, are the only one who can tell us what you don't like about your equipment. Let us know why you don't like the design or performance. Put it on an SF 368 (Quality Deficiency Report). Mail it to us at (insert address of proponent activity). We'll send you a reply.

5. Warranty Information

Some conditions exist which call for special paragraphs. These conditions include paragraphs on warranty, special nomenclature, non-standard abbreviations, and a needed glossary. The following is a typical warranty paragraph which lists the terms and conditions applying to the equipment:

- Warranty Information

  The (insert name of equipment) are warranted by (insert name of company) for (insert time—for example: 12 months or 12,000 miles whichever comes first). It starts on the date found in block 23, DA Form 2408-9, in the logbook. Report all defects in material or workmanship to your supervisor who will take appropriate action through your organizational maintenance shop.

6. Nomenclature Cross-reference List

When, and if, the nomenclature used in the manual deviates from official nomenclature, you need to include a cross-reference paragraph. Nomenclature should be consistent throughout the manual and should define and explain those particular "standard names" used in the manual. Figure 10-1 shows an example of the cross reference list.

7. List of Abbreviations

Any "special abbreviations" and any "critical terms" are also included.

8. Glossary

A glossary of terms to help the user understand a system and a piece of equipment and perform specific tasks should be included when needed.

REFERENCE INFORMATION

This listing includes the nomenclature cross refernce list, list of abbreviations, and explanation of terms (glossary) used in this manual.

A. NOMENCLATURE CROSS REFERENCE LIST

Common Name	Official Nomenclature
loader-Transporter (LT)	launcher, Guided Missile, Carrier Mounted: M752
Launcher (LZL)	Launcher, Zero Length, Guided Missile: M740
Mobility Kit	Handling Unit, Guided Missile Equipment: M39
Fuel Gauge	Liquid Level Indicator

B. LIST OF ABBREVIATIONS

ac	alternating current
amp	ampere
AR	Army regulation

C. GLOSSARY

Hypergolic	Ignite upon contact of components without external aid
Toxic	Poisonous
Ambient	Surrounding on all sides (environment)

Figure 10-1 *Cross Reference Information*

9. Security Classification Marking

Each contract sets specific security requirements. There are standard marking requirements, even when the manual is unclassified. Each company dealing with classified hardware has a security manual and each contract calls out standard markings. You need to read the requirements in each document and know what types of information make the manual classified. You need to follow the prescribed security marking process because this is not an area that accepts "little mistakes." An error here, such as putting classified information in an unclassified manual can have serious effect on national security, and can result in disciplinary action, loss of security clearance and possible loss of employment. Failure to safeguard classified source data can have the same impact.

Section II   Equipment Description

The Equipment Description section contains the following information:

1. Equipment Characteristics, Capabilities, and Features
2. Location and Description of Major Components
3. Differences Between Models
4. Equipment Data

Section II provides the user with a functional description of the equipment operation. It may give the user an overview of what the controls do, but it does not go into detail. A description of the controls and what to expect in response to control actions is described in the operation chapter. It

MIL-M-63036A(TM)

Purpose of TOW Weapon System

A crew portable, heavy antitank weapon designed to attack and defeat armored vehicles and other targets such as field fortifications.

Capabilities and Featuures

Major weapon system components:

a. Launcher
   1. Launcher tube
   2. Traversing unit
   3. Missile guidance set
   4. Battery assemblies (2)
   5. Optical sight
   6. Tripod

b. Missile (tube-launched, optically tracked, wire-command-link)
   All weather operational
   High first-round his probability
   Highly portable
   Built-in self-test circuits

Figure 10-2 Concise Reference Information

also gives a top-down, top-level description of the system function, but only to the level the users need to know for safe and correct operation.

1. Equipment Characteristics, Capabilities, and Features

The hardware is described in the equipment characteristics section, but the information is restricted to abbreviated

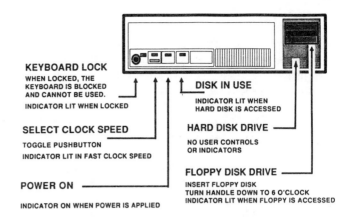

KEYBOARD LOCK
WHEN LOCKED, THE KEYBOARD IS BLOCKED AND CANNOT BE USED.
INDICATOR LIT WHEN LOCKED

SELECT CLOCK SPEED
TOGGLE PUSHBUTTON
INDICATOR LIT IN FAST CLOCK SPEED

POWER ON
INDICATOR ON WHEN POWER IS APPLIED

DISK IN USE
INDICATOR LIT WHEN HARD DISK IS ACCESSED

HARD DISK DRIVE
NO USER CONTROLS OR INDICATORS

FLOPPY DISK DRIVE
INSERT FLOPPY DISK
TURN HANDLE DOWN TO 6 O'CLOCK
INDICATOR LIT WHEN FLOPPY IS ACCESSED

Figure 10-3 Combined Clear Words and Simple Pictures

descriptions of the characteristics, general capabilities, special features, and limitations of the system. You show the type of equipment, portability or mobility, special operating and environmental features, and remote control features. But you do not describe the major components yet because they will be described in the paragraphs which follow. You use tables, lists, and illustrations and avoid words that force the user to use a glossary. In addition, common abbreviations (MIL-STD-12) and concise reference information are used. Figure 10-2, taken from MIL-M-63036A (TM), is an example.

MIL-M-63038B (TM)

1-6. DIFFERENCES BETWEEN MODELS

	M113A1	M577A1	M106A1	M125A1	M132A1	XM741
CARRIER FUNCTION						
Personnel/cargo	X	---	---	—	—	---
Command post	---	X	---	—	---	---
Mortar carrier	---	—	X	X	---	---
Flame thrower	---	—	—	—	X	—
Anti-aircraft gun carrier	---	—	—	—	—	X
ARMAMENT AND FIRE-CONTROL						
M108 flame thrower	—	—	—	—	X	—
107-mm mortar	—	—	X	—	—	—
81-mm mortar	—	—	—	X	—	—
Caliber .50 machine gun	X	—	X	X	---	---
7.62-mm machine gun	—	—	—	—	X	---
AUXILIARY EQUIPMENT						
Exhaust extension	—	—	X	X	—	X
Air grille curtain	—	—	X	X	—	X
Covered extension	—	X	—	—	—	---
Generator set/cover	—	X	—	—	---	—
Auxiliary power unit	—	—	—	—	—	X
Electronic heater kit	—	X	—	—	---	---
Personnel heater kit	X	X	X	X	---	---
Engine coolant heater kit	X	X	X	X	—	---
Driver's windshield kit	X	X	X	X	—	X
Litter kit	X	—	—	—	—	—

Figure 10-4 Tabulate the Difference Between Models

2. Location and Description of Major Components

Both internal and external pictures of the system are used to show the general features, essential operating items, and each major component and accessory. Callout numbers or block text on the illustrations link to keyed text descriptions. You provide only a physical description of the items and describe them from the user's point of view, excluding any information that will be discussed in the "Equipment Data" section which comes later. Figure 10-3 shows a clear, well-planned illustration. Note that the equipment is shown pictorially and explained clearly. The specification applied may allow use of clear text or identification of each component by a "call-out" number at the end of the pointer.

You combine clear, planned pictures with well-chosen words to describe the hardware and identify key components for the user. Tell the users what they need to know to use the system safely—to prevent injury to people or damage to the system.

3. Differences Between Models

The "difference" paragraph is needed to describe any significant differences between models or between units of the same model which affect the user's actions. You indicate the extent of "interchangeablility" and relate these differences explicitly to equipment model number, part number, or range of serial numbers so that the users can clearly identify any equipment configuration. If model or unit differences exist but do not affect operation or maintenance

TYPICAL EQUIPMENT SPECIFICATION	TYPE OF EQUIPMENT						
	Hand & Portable Weapons	Combat Vehicles	Trucks	Construction & Material Handling Equipment	Generator Sets	Comm & Electronics	Test and Diagnostic Equipment and Audio Visual Eq
Caliber of Weapons	•	•					
Capacities [cooling, fuel, oil, stowage, ammunition, etc.]	•	•	•	•	•		
Center of Gravity	•	•	•	•	•		
Clearance	•	•	•	•			
Detection Characteristics [frequency, range rate, etc.]	•	•					
Dimensions [emplacement, travel, overall, etc.]	•	•				•	•
Engine Characteristics [bore, cylinder, rpm, etc.]		•	•	•	•		
Field of Fire	•	•					
Field of View		•					
Gear Speed		•					
Ground Pressure	•	•	•	•			
Land Performance [speed, towed speed, grade, turning radius, etc.]	•	•	•	•			
Loads [maximum]	•	•	•	•			
Make, Model, Equipment Type, Major Components	•	•	•	•	•	•	•
Muzzle Velocity, Pressure	•	•					
Operating Modes	•	•	•	•	•	•	•
Operating Temperatures	•	•		•	•	•	•
Optical Characteristics	•	•					
Range of Fire	•	•					
Rates of Fire	•	•					
Rifling	•	•					
Slow Rate	•	•					
Tire Pressure	•	•	•	•			
Water Performance	•	•	•	•			
Weights [component, overall, bridge classification, etc.]	•	•	•	•	•	•	•
Accuracy [tolerance]	•	•	•	•	•		
Antenna Characteristics						•	•
Band Width						•	•
Band Sets						•	•
Calibration Requirements						•	•

Figure 10-5 *Some of the Equipment Data Needed*

actions, you have to explain this. Figure 10-4, taken from MIL-M-63038B (TM), shows a typical example of the "difference table."

4. Equipment Data

You need to provide equipment data, including numerical and other specification related data which includes top functions and summarizes specific capabilities and limitations of the system. You also summarize other operation and maintenance data. Figure 10-5 shows a typical Equipment Data Table.

Section III Technical Principles of Operation

The Technical Principles of Operation section contains a functional description of the equipment operation aimed at the depth needed by the user for safe use and for understanding the relationship between equipment operation and maintenance. This section tells the user by use of functional descriptions how the controls and indicators interact with the rest of the system. You emphasize the user's functional

needs but you do not emphasize the highly technical principles of operation for those parts of the system that the user is not expected to encounter nor needs to know about. Your readers must be able to understand how it operates so that when it breaks down or ceases to function properly, they can perform the necessary maintenance. Figure 10-6 shows a relatively complex mechanical system and the depth of information typically needed for the user/operator.

Writing Style

When writing the Description Chapter you need to use a consistent, parallel descriptive pattern for each section. The introductory chapter describes the system from the top down, from the outside in, but most importantly from the user's point of view. You need to answer the following questions in this first chapter:

- What does the system do?
- What are the major components?
- How do the components fit into the whole system?

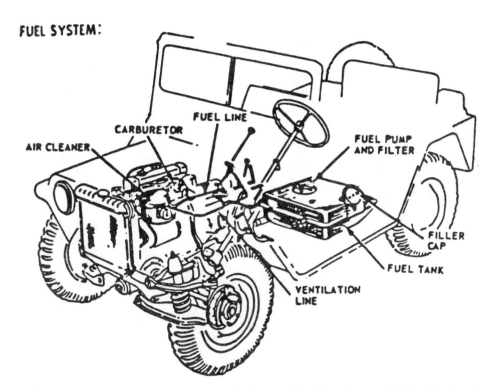

FUEL SYSTEM:

Carburetor — is a side draft, single barrel type. It is mechanically controlled with accelerator pedal, hand throttle and manual choke. For underwater operation it is vented to the air cleaner.

Air Cleaner — is an oil bath type. During underwater operation air intake pipe must be attached to the air cleaner.

Fuel Pump and Filter — are located in the fuel tank. The pump is a plunger type operated electrically that ticks when it is in operation. All fuel must pass through the filter element before it enters the pump.

Fuel Tank — houses the electric fuel pump and fuel filter. Nos. 17.7 gal (67 liters) capacity and a drain plug in bottom center of tank. A fuel strainer is located in the tank filter neck.

Filler Cap — is pressure type, includes a vent valve to seal the vent during fording only.

Connecting Lines — carry fuel from the tank to the carburetor and provide tank venting.

Figure 10-6 *A Typical Set of "Principles of Operation" Explained at the Operator's Level of Need*

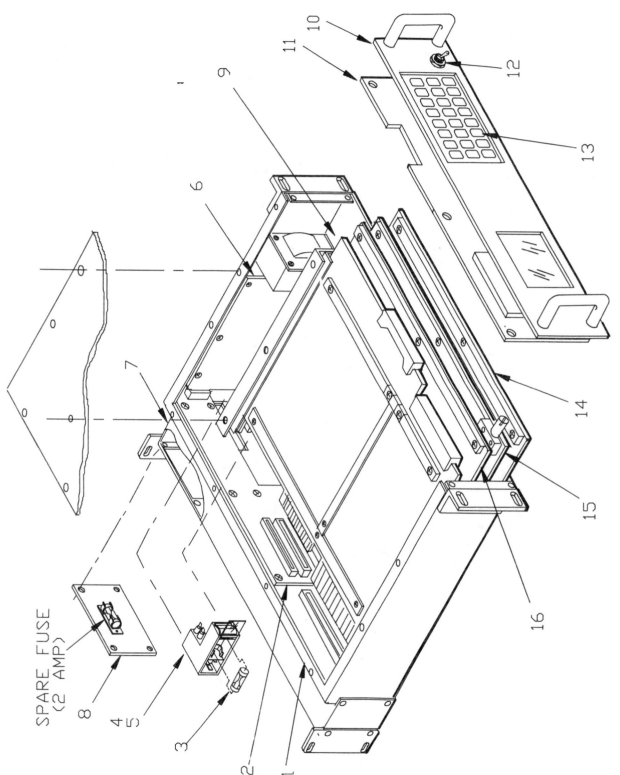

SPARE FUSE
(2 AMP)

Figure 10-7 *Example of Callouts*

The following are the most common organizational patterns to use when giving a brief description of each major component:

- Spatial
- Cause to Effect
- Functional

Spatial descriptions may be organized in a horizontal, vertical, or circular pattern. A horizontal pattern leads the user from a "front-to-rear" or from a "side-to-side" view of the system. A vertical pattern leads the user from a "top-down" on from a "bottom-up" view of the system. A circular pattern leads the user from an "around the clock" view. Sometimes, the callout number is used, starting at about "12 o'clock" and continuing in a clockwise rotation to indicate all parts (Figure 10-7). The spatial pattern provides the simplest descriptive view because you describe the system/hardware/equipment the way it "looks or lies in space."

Cause-to-effect descriptions work better when describing controls and indicators. You explain what action causes what function in the system. You may need to combine both spatial and cause-to-effect organizational patterns to make your points clearer to the user; however, you will probably use more cause-to-effect patterns in chapters that come later in the manual. Functional descriptions tell "how the system works."

Conclusion

In the Description Chapter, you tell the reader about both the manual and the equipment. Some of the information included in the Description Chapter of the O&M Manual is specified standardized "boilerplate" text, illustrations, and precaution messages. In the rest of the chapter you describe the manual and the "top-level" functions (explaining how the system works) by combining simple functional block diagrams with functional descriptions and with tables of the system's key characteristics. Both a clear writing style and the compliance with specific format and content are necessary to create a useful manual.

Chapter 11

The Installation Chapter in the Manual

Introduction

The Installation Chapter provides the installers and maintenance personnel with all the information they need to insure that the system is adequately inspected, serviced, and tested before it is put into normal use. Generally, this chapter does not duplicate information found in other sections of the manual. This chapter tells the users all they need to know in order to do the following:

- Unpack the system
- Assemble the system
- Inspect the units for damage
- Understand the special needs of the system, such as
  cooling
  plumbing
  facilities
  foundation
  power
  clearance
- Connect the system to power, ground and signal interfaces
- Fasten the system in place
- Repack the system or units of the system

You use illustrations, sketches, tables, and text to tell users how to unpack and install the system. Within the limits of the customer's specified format and contents, you plan a clear, simple process description, so that the users will not have these typical complaints:

"I've finished putting it together and I have parts left over."

"I can't find the step that tells me how to get these together."

Sections in the Chapter

The procedures must lead the users logically from beginning to end with no ambiguity and with no missing steps. If the system is "complex," you may need to use some of these divisions in the Installation Chapter:

Section I   Site and Shelter Requirements
Section II   Service Upon Receipt of Material
Section III   Installation Instructions
Section IV   Preliminary Servicing and Adjustment of Equipment
Section V   Circuit Alignment

You do not repeat material, but you may need to reference other applicable sections of the manual when needed.

Section I   Site and Shelter Requirements

The Site and Shelter Requirements section tell the installers and maintenance personnel about the physical needs peculiar to this system (i.e. site and shelter selections). Anticipate questions such as:

- How much space is needed?
- How much cooling is needed?
- How flat must the site be?
- How clear must the site be (of trees, buildings, obstructions, etc.)?

1. Siting Requirements

The Siting Requirements paragraph tells readers only what they need to know to avoid damage, injury, or unnecessary delays, or to identify operational limitations. Siting requirements include the following:

- Location
- Proximity to power sources
- Effective range
- Terrain requirements for effective use
- Technical requirements
- Shelter locations
- Acceptable site plans or layouts

When applicable, instructions should be given to help installers overcome adverse siting conditions. If the system includes elements which must be oriented to some baseline, include the orienting procedures.

2. Shelter Requirements

Whenever the system is to be housed in a building or some other semipermanent shelter, the Shelter Requirements paragraph stipulates that you include the following:

- Amount of floor, wall and vertical space needed
- Plan (or acceptable alternate plans) for equipment location
- Needed weight capacity of the supporting floor
- Dimensions of the installed equipment
- Total weight the floor must support and the area of the floor that is actually supporting the weight

• Environmental conditions needed (power, ground, cooling, air, etc.)

You do not include architectural or engineering information that would confuse the installers and maintenance personnel, but you do show, both in the floor plans and in the equipment elevations, what the space requirements for storage of support equipment, supplies, spares, and space are needed for proper maintenance.

Section II   Service Upon Receipt of Material

The Service Upon Receipt of Material section contains instructions that users must follow to unpack, inspect, and deprocess the equipment. You must keep the instructions as simple and unambiguous as possible and use illustrations and tables as needed. Figure 11-1, taken from MIL-M-63038B (TM), shows a typical format required by the government.

1. Unpacking

You should omit the unpacking paragraph if no special sequence of actions is needed or if such instructions will not result in greater efficiency during the unpacking phase. For example, units which are to be installed in an equipment rack may not need to be unpacked until the rack is unpacked

and installed. This paragraph may contain only a statement such as the following:

There are no special unpacking considerations; however, save all the boxes and packing materials for reuse when system is removed or components are returned for repair.

When more detailed instructions are needed, they must be step-by-step procedures that are keyed to illustrations.

2. Inspection and Verification of Received Equipment

The customer specification may define verbatim content for the Inspection paragraph. For example, MIL-M-63038 (TM) requires these statements:

• Inspect the equipment for damage incurred during shipment. If the equipment has been damaged, report the damage on DD Form 6 Packing Improvement Report.
• Check the equipment against the packing slip to see if the shipment is complete. Report all discrepancies in accordance with the instructions of TM 38-750.
• Check to see whether the equipment has been modified.

These examples of statements required by the government do not tell the users "how to" inspect or how to check, only what to do. There is no restriction against adding to the required information data that will explain to the users just what to look for when they inspect the equipment and

MIL-M-63038B (TM)

2-3. SERVICE UPON RECEIPT — M29 and M30 CONTROL SURFACES AND CONTAINERS			
LOCATION	**ITEM**	**ACTION**	**REMARKS**
1. Container	Components	a. Inspect for rust, fungus, paint damage and deformation.	Para 4-2
		b. Reject container if damage prevents it from functioning properly.	- - -
2. M29	Control surfaces	a. Inspect for dents and scratches on post, trailing edge phenolic, skin, and closure plate.	Para 4-3
		b. Reject control surface.	- - -
		(1) If post dents or scratches exceed 0.002 in.	
		(2) If trailing edge phenolic dents exceed 0.040 in.	
		(3) If skin dents exceed 0.030 in. within 2 in. of post.	
		(4) If closure plate dents exceed 0.030 in. within 2 in. of post.	
3. M30	Control surfaces	a. Inspect for dents and scratches on post and skin.	Para 4-18
		b. Skin dents or scratches up to 0.050 in. are allowable, but should be blended.	- - -
		c. Reject control surface if post dents or scratches exceed 0.002 in.	- - -

Figure 11-1 *Service on Receipt Tabular Format*

what to do to check the equipment. Most of all, you must clearly show them how to determine if the "equipment has been modified," so they may use appropriate variations of the procedures for the specific configuration of equipment.

3. Deprocessing Unpacked Equipment

Many of the equipment items that users are unpacking have been "processed" for transportation. Some equipment are packed in heavy lubricant or preservative agents. Electromechanical assemblies (i.e. tape decks, etc.) have shipping clamps or other shipping devices installed. Hard disk drives may be shipped with heads packed. In this "deprocessing" paragraph, you tell users what to do to convert the items from the shipping condition to a usable condition. If possible, refer to other sections of the manual to avoid repeating instructions. If another manual exists that describes the actual process and that manual is known to be available to users, you can refer to it. When including detailed deprocessing instructions, you should be clear and concise and tell users what they need in order to do the tasks.

Section III   Installation Instructions

When providing an overview of the Installation Instructions section, give users all the information needed to install the system elements properly in order to make the necessary interconnections and to do the initial lubrication and adjustments. There are usually four paragraphs:

1. Tools, Test Equipment and Materials Required for Installation
2. Assembly of Equipment
3. Installation
4. Interconnections

1. Tools, Test Equipment, and Materials Required for Installation

You list each tool, model of test equipment, and all materials and supplies needed to do the installation tasks in this paragraph. Show the name, part number, and the quantity of each item that is listed. The items listed must be available to the installers either as part of the system or part of the organization's authorized property list. When essential for proper or safe use, illustrate the correct use of the item. Include properly formatted warnings when needed.

2. Assembly of Equipment

This paragraph contains detailed, step-by-step instructions for assembling a system which was shipped disassembled. Illustrate the steps if needed for clarity and include the instructions needed to assemble equipment racks. Tell users how to mount units into racks properly and how to fasten the racks for safe operation. When heavy equipment is to be installed, it is imperative that proper warnings and cautions related to human and equipment safety be inserted in the procedure prior to the step describing the lifting action.

3. Installation

This paragraph provides step-by-step instructions with the necessary illustrations and includes each action (place, mount, attach) needed to install cable and wiring connections and to use tools and test equipment properly (Figure 11-2). On the accompanying illustrations, show all dimensions which must be maintained when placing, mounting, or attaching items. You place the steps in the sequence which allows the most efficient, quickest, most accurate installation. If initial adjustments or orientation of items is best done while installing the system, show those steps in the proper sequence but do not include instructions for connecting the system to external cables at this time. If the system is designed for use in different types of installation, you show clearly the proper process for each type. If the installers need help from maintenance personnel who have higher skill levels and training, include a note such as the following:

NOTE

The following installation procedure must be made with assistance of (insert category) maintenance personnel (include skill code if needed).

Installation instructions are complete ONLY when they include instructions for all the following:

1. all required options
2. accessory items
3. auxiliary items (to extend or increase system capability)
4. grounding of system for safety and for proper operation
5. torque requirements and other additional task requirements

4. Interconnections

In this paragraph, you must include clear instructions for connecting all cables, pipes, hoses, air lines, etc. needed by the system/hardware/equipment. Interconnection diagrams are one of the better ways to show this information. Include routing instructions for each alternative site layout when applicable. These interconnection instructions usually include all system cabling (rack to rack, unit to unit, and system to host site and other systems). If the system is so complex that a separate interconnection volume or appendix is needed, you refer to that document in this paragraph. The document must clearly identify each cable, each mounted connector to which a cable will be mated, and the desired route for each cable.

You must provide special instructions or special procedures needed for safe mating of connector pairs. A diagram which identifies individual wires and color code or wire identification of the terminal at each end of the wire may be required if installation requires connection of wires as well as cables. You explain alternate connection patterns for site variation or operation mode variation. If a wiring pattern is repeated (replicated) within the system, show it only once and clearly explain where the pattern occurs. If

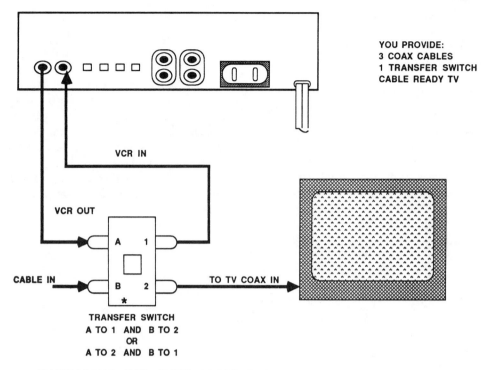

YOU PROVIDE:
3 COAX CABLES
1 TRANSFER SWITCH
CABLE READY TV

VCR IN

VCR OUT

CABLE IN

A   1

B   2
*

TO TV COAX IN

TRANSFER SWITCH
A TO 1 AND B TO 2
OR
A TO 2 AND B TO 1

INSTALLING VCR WITH CABLE TRANSFER SWITCH

Figure 11-2 *Typical Installation Illustration*

controls must be preset before connection is made, you show the control pattern clearly.

5. Installing "Plug-In" Items

When the installation includes insertion of circuit card assemblies (CCA), modules, or other separately packed items, you must give step-by-step procedures for handling, inspecting, and installing of the items in this paragraph. You would reference the installation process in the Maintenance Chapter for further details.

6. Special Applications

The special application paragraph identifies those special actions that apply to a class of similar installations such as mounting a radio unit on a vehicle, or on a boat, or in an aircraft. You show the class considerations, not the special or peculiar considerations, for installing the unit on one particular vehicle.

7. Van and Shelter Installations

To the extent needed for maintenance, you show the installation and removal process for each non-permanent unit mounted in a van or shelter. In this paragraph, you include diagrams and drawings of all electrical and signal wiring interconnections and include cable runs, equipment location, and circuit breaker panels.

Section IV   Preliminary Servicing and Adjusting of Equipment

The Preliminary Servicing and Adjustment of Equipment section includes instructions for each service action (lubrication, filling reservoirs, etc.) and adjustment action needed to make the system ready for operation. You put any needed precautionary alert symbols (warning, caution, note) where users need them for safe accomplishment of these tasks. You include needed illustrations and you reference other chapters in the manual for details to avoid redundancy. Provide instructions for making those tests, adjustments, and servicing actions needed. Typical actions may include:

- Verify correct interconnections.
- Ensure adequate clearance for moving elements.
- Verify setting of each control.
- Ensure that power may be applied safely, without damage to persons or to equipment.

Inspect or check (test) all items which are needed and are reasonable and practical. The list below is an example:

- Automatic controls adjusted
- Covers, doors, plates in place
- Liquid coolers operating
- "Burn-in" accomplished
- Installer controls set properly
- Operator controls set properly

- Operator tests normal
- Preliminary test passed
- Proper cable routing
- Proper grounding
- Proper "strapping"
- Safety interlocks and switches

- Plug-in parts firmly connected
- Prevent equipment damage
- Proper connections—wires, etc.
- Proper "orientation"
- Proper "warm-up" stabilization
- Ventilation and proper cooling

Refer to the operation paragraphs for operator control settings, warm-up, or stabilization times and refer to the installation paragraphs for installation control settings, orientation, strapping, and burn-in times.

Section V   Circuit Alignment

The Circuit Alignment section includes instructions for the circuit alignment procedures needed during installation, and shows external connections, switch settings, patch panel connections, internal control settings/strapping needed, and alignment procedures. Use illustrations to help clarify the instructions and reference other chapters in the manual as needed to avoid redundancy.

1. External Connections

This paragraph tells users which external lines must be connected to the system before alignment and operation, and uses step-by-step instructions for the connection tasks as you did in the installation wiring/cabling instructions.

MIL-HDBK-63038-1A (TM)

2-46.   REPLACE DRIVE SHAFT (cont)

__INSPECTION__

INSPECT DRIVE SHAFT.

a. Visually inspect new drive shaft (4) for bends and cracks.

b. Get a new drive shaft if above conditions are found. Repeat steps a. and b.

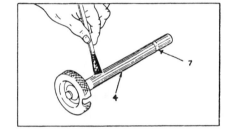

__LUBRICATION__

LUBRICATE DRIVE SHAFT.

Apply light coat of grease to entire length of drive shaft (4) splines.

__INSTALLATION__

1. INSTALL DRIVE SHAFT.

a. Insert small end of drive shaft (4) in opening in bottom of receiver assembly.

b. Push drive shaft (4) in as far as it will go. Push button (5) in center of drive shaft handle (6). Release button.

c. Pull on drive shaft handle (6) to make sure that drive shaft is locked in place.

d. Use retaining ring pliers. Install retaining ring (3) in groove (7) on drive shaft (4).

2. PERFORM RECEIVER MANUAL CYCLE CHECK (para 2-10).

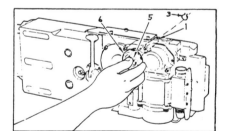

NOTE

FOLLOW-ON MAINTENANCE:
Install receiver (gun)
assembly (para 2-42).

Figure 11-3 *Typical Installation Instructions*

Show all external connections that apply specifically to the alignment process.

2. Switch Settings, Patch Panel Connections, Internal Control Settings

This paragraph tells users clearly, using illustrations as needed, where to set each control. In addition, connection lists show how to connect each patch panel. Identify each internal control setting and how each "strapping" should be connected, giving this information for each alignment and for each mode of operation needed.

3. Alignment Procedures

This paragraph provides step-by-step instructions for each alignment needed and shows users how to do each alignment in each mode that applies during installation. If possible, refer to the alignment process descriptions in the Maintenance Chapter in the manual.

Writing Style

When writing the Installation Chapter you need to give users a general overview: where they start, what they need, what they need to do, and what to expect when they do it properly. Form each procedural step in the command mood (imperative) and start each action step with an action word (verb). You show conditional actions by using conditional words (may, could, might, etc.). Clearly "locate," "name," and "describe" each step to the installer.

Detailed instructions often need illustrations or remarks to help users learn what they need to know. You may need to write internal directions if the procedure "branches" or jumps over steps. Figure 11-3, taken from MIL-HDBK-63038-1A (TM), shows an example of a simple installation instruction, yet it does have the following flaws:

- The figure does not specify the size of the cross-tip screwdriver.
- The steps shown do not start with an action word.
- The action words chosen are not clear and even are ambiguous.

Each detailed procedure must be complete: completely accurate and completely reliable.

Conclusion

The Installation Chapter leads the reader through the critical process needed to unpack, install, and start using the equipment. Tell users what to expect when the process is done correctly by using check lists, control setting lists, illustrations, etc. You need to refer to tools and equipment and supplies that are familiar to the users or include definitions. There are more problems created by using improper installation techniques than there are in the equipment design. When you create an effective Installation Chapter you help minimize the potential for failures caused by the operator.

Chapter 12

The Operation Chapter in the Manual

Introduction

The Operation Chapter tells the users and operators how to operate the system/hardware/equipment. It includes information on the following:

WHERE:	the controls and indicators are
WHAT:	each control does and what is ''normal'' for each indicator
WHICH:	mode to use in each set of conditions
WHEN:	to use each mode
WHY:	to use each mode
HOW:	to operate in an emergency, to TURN ON, CHECK OUT, and SHUT DOWN the system

This chapter discusses equipment operation without providing detailed mission or tasks procedures. For example, you would tell a radio operator how to turn the radio on, how to select a frequency and, in general, how to use the radio. You would not tell the operator what to say or what operating protocol to use. Other manuals or resources may be referred to for discussion of the mission or task processes. In very small manuals, you may combine this ''how to'' chapter with the Theory of Operation Chapter. You have already told users/operators what the system is in the Introduction/Description Chapter and where to put it and how to install and connect it in the Installation Chapter. Now you need to give a clear picture that will allow users to use the system in any expected set of conditions.

Clearly describe each control and each indicator operators may need to use and accurately tell them how to activate or start-up the system. Tell them what tests to do to verify that the system is ready for use. Explain each possible mode of operation, under both USUAL and UNUSUAL conditions. You also clearly explain how to use the system in an emergency.

Sections in Chapter

Each manual specification defines what information you must present in this chapter. MIL-M-63036 (TM) specifies four sections for the Operation chapter. Each section starts with a general introductory paragraph which tells the operators (1) what the section describes, (2) what they should learn when they read the section and (3) what skills they should have when they are familiar with the contents of the section. The four sections are as follows:

Section I	Description and Use of Operator Controls and Indicators
Section II	Preventive Maintenance Checks and Services
Section III	Operation Under Usual Conditions
Section IV	Operation Under Unusual Conditions

Section I   Description and Use of Operator Controls and Indicators

The Controls and Indicators section describes each control and each indicator the operator will use. You plan each illustration to show the controls and indicators as seen from the operator's point of view. Label the illustrations to help the operator understand what the controls are, and use the exact name of the control as it is on the panel. If the panel has no control name on it, you assign a simple, clear, functional name for the control and then consistently use the name in the rest of the chapter. If needed, include such functional names in the glossary at the end of the manual.

Figure 12-1 has an effective mix of words and illustrations to explain controls and indicators. For very simple items, you may combine the operating instructions with the explanation of principles of operation. Figure 12-2 shows an example of such a combined illustration. If there are numerous controls and indicators as in a complex system, you may find that a table such as the one in Figure 12-3 communicates better.

Section II   Preventive Maintenance Checks and Services

The Preventive Maintenance Checks and Services (PMCS) section tells the operators which preventive maintenance checks and services must be done to the system. Preventive Maintenance (PM) is a class of actions taken to keep the system operating the way it should. PM includes tests, adjustments, inspections, cleanings, lubrications, and occasionally a ''hard-time'' replacement. Service is a subclass of PM. Service involves resupply, lubrication, and ''topping up'' actions; a familiar example of PM is ''washing the car,'' while a familiar service is ''filling the gas tank.'' The PMCS section usually has three parts:

- PMCS Introductory Material
- PMCS Special Instructions
- PMCS General Requirements

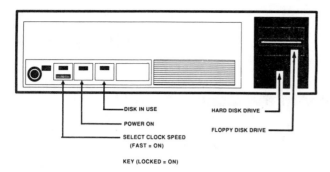

DISK IN USE

POWER ON

SELECT CLOCK SPEED
(FAST = ON)

KEY (LOCKED = ON)

HARD DISK DRIVE

FLOPPY DISK DRIVE

Figure 12-1   *Description of Controls and Indicators*

The PMCS tasks are drawn from such sources as the Logistics Support Analysis Record (LSAR), vendor manuals, and the writer's experience. They are discussed in detail in the Maintenance Chapter of the manual.

1. PMCS Introductory Material

This paragraph includes statements which tell the operators what to remember before they operate, while they operate, and after they operate the system. It also tells them what to remember when the system fails to operate and when the system fails to operate correctly. After these ALERT statements, you introduce the operator to the PMCS needed and tell them what they are expected to do, why they do

it, and what to do if they find an indicator showing an abnormal indication. You explain the headings in the PMCS tables and tell operators how to report discrepancies and problems they find during the PMCS. Tell them how to report discrepancies and problems they find in the manual and tell them where in the manual to find instructions for disassembly and assembly of the unit when the PMCS requires those actions.

2. PMCS Special Instructions

This paragraph gives the operators those "special" instructions needed to do the PMCS correctly. In many cases, these special instructions are defined by the manual specifications. For example, if the PMCS must be done on operating equipment, MIL-M-63036 (TM) requires this statement between the PMCS chart title line and the chart itself:

NOTE

If the equipment must be kept in continuous operation, check and service only those items that can be checked and serviced without disturbing operation. Make complete checks and services when the equipment can be shut down.

At times, the general PMCS instructions must change to match the conditions. Systems are not always used in just one environment, so the changed operating environment may indicate changes in the period between PMCS or may

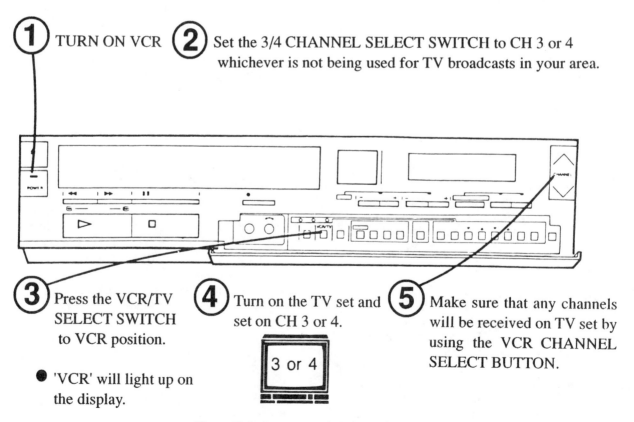

1 TURN ON VCR   **2** Set the 3/4 CHANNEL SELECT SWITCH to CH 3 or 4 whichever is not being used for TV broadcasts in your area.

3 Press the VCR/TV SELECT SWITCH to VCR position.

● 'VCR' will light up on the display.

4 Turn on the TV set and set on CH 3 or 4.

3 or 4

5 Make sure that any channels will be received on TV set by using the VCR CHANNEL SELECT BUTTON.

Figure 12-2   *Description of Mechanical Controls*

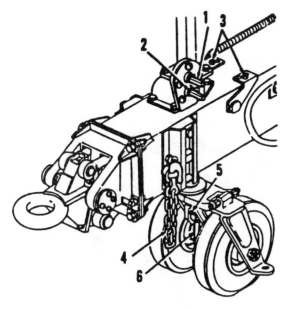

Key Control or Indicator	Function
1 Cranking stud (High speed)	Extends or retracts caster jack
2 Cranking stud (low speed)	Extends or retracts caster jack and allows 2-man operation with missile round on board.
3 Caster up lock stud (2)	Engages with hole in lug on caster forks to secure caster wheels in stowed position.
4 Tow bar safety chain (2) shown in stowed position	Provides a safety factor when the tow bar is attached to a towing vehicle.
5 Recess for pry bar	Facilitates steering of caster wheels on difficult terrain.
6 Caster down lockpin	Locks caster wheels in down position.

Figure 12-3 *Tabular Method for Description of Controls and Indicators*

change what PMCS actions are needed. For example, an air filter will have to be cleaned more often in the Sahara Desert than in mid-Ohio (changed period) and the operator will have to test the effectiveness of anti-freeze in a radiator in cold climates but not in hot climates (changed need). Most manual specifications allow footnotes in the PMCS table to show these deviations. Another use of special instructions involves changes to the PMCS schedule because of the number of operators or the infrequent use of the system. You use statements such as those in Figure 12-4 before each PMCS table. (See also Figures 12-5, 12-6, 12-7 taken from MIL-HDBK-63038.)

3. PMCS General Requirements

This paragraph tells the operators what to do in general. The PMCS tasks are selected by engineers using some form

Perform weekly as well as before operations PMCS if:

1. You are the assigned operator and have not operated the item since the last weekly PMCS.

2. You are operating the item for the first time.

or

Leakage definition for operator/crew PMCS shall be classified as follows:

Class I	Seepage of fluid (as indicated by wetness or discoloration) not great enough to form drops.
Class II	Leakage of fluid great enough to form drops but not enough to cause drops to drip from item being checked or inspected.
Class III	Leakage of fluid great enough to form drops that fall from the item being checked/inspected.

CAUTION

Equipment operation is allowable with minor leakages (Class I or II). Of course, you must consider the fluid capacity in the item/system being checked/inspected. When in doubt, notify your supervisor.

When operating with Class I or Class II leaks, continue to check fluid levels as required in your PMCS.

Class III leaks should be reported to your supervisor or organizational maintenance.

Figure 12-4 *Typical Prefacing Notes to PMCS*

of logical decision analysis. One of the more effective is a decision tree method called Reliability Centered Maintenance (RCM) which was developed by the commercial airline industry and adapted by the military. Your information for these PMCS tables should come from the LSAR, but you may have to dig it out by interviews with the design engineers or the maintenance engineers. You arrange the PMCS tables in a logical order and design a sequence which optimizes time, motion, interference, and operating restrictions. Many PMCS tables use column headings such as the following:

- Equipment is not Ready/Available if . . .
- Item Number
- Item Inspected
- Interval
- Procedure

Section III Operation Under Usual Conditions

The Operation Under Usual Conditions section contains step-by-step instructions for all actions needed to operate the system in normal conditions. You explain any combination of control settings which might damage the system or endanger the operators or other persons. You include appropriate alert messages (Warning, Caution, Note) so the operators will be aware of potential damage or trauma. It is best to use a time sequence to order your instructions. Remember that step-by-step instructions include the following:

Table 1. QUARTERLY PREVENTIVE MAINTENANCE CHECKS AND SERVICES

SEQUENCE-ITEM	PROCEDURE
52 Gun Tube	Clean and lubricate. Check for unusual wear, erosion, or damage in the bore.
53 Torque Key	Remove, measure, and lubricate torque key as follows:
	A Remove lock wire (1).
	B Use a 5/16-inch socket key to remove ten screws (2) and washers.
	C Lift key out, measure width of keyslide (3) using a mechanic's scale with 1/32-inch graduations. Width should be at least 31/32 inches (24.6-mm). If not, replace key. Lubricate and reinstall.
	NOTE
	Key wears approximately 0.001 inch per 100 rounds fired.
54 Breech Ring and Block (to include all components)	Inspect for cracks, burrs, and ease of operation. Clean and lubricate as follows:
	A BREECHBLOCK AND CARRIER DETENT PLUNGER. Make sure breechblock rotates fully. Check smoothness of opening and closing. Lubricate threads (1) of breech ring and threads (2) of breechblock. Inspect carrier detent plunger (3) for distortion and wear. If slides of plunger are deformed or noticeably worn, get a replacement.
	B OBTURATOR SPINDLE. Check vent hole (4). Clean by pushing vent cleaning tool (5) through vent and rotate clockwise until hole is clean. Next, run cleaning brush (6) through several times. Remove any traces of powder with rifle cleaner. Wipe dry and lubricate.
	C OBTURATOR PAD. Check obturator pad (7) for damage. Clean with soap and warm water and dry. Check to insure split rings are 180 degrees apart when assembled.

Figure 12-5   Example of a Preventive Maintenance Format

Table 0-0 Organizational Preventive Maintenance Checks and Services

| W-Weekly | Q-Quarterly | A-Annually | H-Hours |
| M-Monthly | S-Semiannually | B-Biennially | MI-Miles |

Item No.	Interval								Item To Be Inspected	Procedures
	W	M	Q	S	A	B	H	MI		
1	•								XXXXXXXXXX	XXXXXXXXXXXXXXXXXXXXXXXXXXXXXXXXXXXXXXXXXXX XXXXXXXXXXXXXXXXXXXXXXXXX
2			•						XXXXXXXXXX	XXXXXXXXXXXXXXXXXXXXX
3						•			XXXXXXXXXX	XXXXXXXXXXXXXXXXXXXXXXXXXXXXXXXXXXXXXXXXXXXX XXXXXXXXXXXXXXXXXXXXXXXXXXXXXXXXXXXX
4							50		XXXXXXXXXX	XXXXXXXXXXXXXXXXXXXXXXXXXXXXXXXXXXXXXXXXXXXX
5								5000	XXXXXXXXXX	XXXXXXXXXXXXXXXXXXXXXXXXXXXXXXXXXXXXXXXXXX
6					•				XXXXXXXXXX	XXXXXXXXXXXXXXXXXXXXXXXXXXXXXXXXXXXXXXXXXXXX XXXXXXXXXXXXXXXXXXXXXXXXXXXXXX
7		•							XXXXXXXXXX	XXXXXXXXXXXXXXXXXXXXXXXXXXXXXXXXXXXXXXXXXXXX
8				•					XXXXXXXXXX	XXXXXXXXXXXXXXXXXXXXXXXXXXXXXXXXXXXXXXXXX
9		•							XXXXXXXXXX	XXXXXXXXXXXXXXXXXXXXXXXXXXXXXXXXXX
10	•								XXXXXXXXXX	XXXXXXXXXXXXXXXXXXXXXXXXXXXXXXXXXXXXXXXX

Figure 12-6 Another Preventive Maintenance Format

a. Item number column. Checks and services shall be numbered in chronological order regardless of interval, unless otherwise specified by the procuring activity. An explanation shall be included that this column shall be used as a source of items numbers for the ''TM Number'' column of DA Form 2404, Equipment Inspection Maintenance Worksheet, in recording results of PMCS.

b. Interval colums. The columns headed ''B'', ''D'', ''A'', ''W'' and ''M'' shall contain a dot (.) opposite the appropriate check. Thus, if a given check is performed before operation a dot shall be placed in the ''B'' column; if the check is accomplished during operation the dot should be placed in the column headed ''D'' and, if the same check is made in two or more periods, a dot will be placed in each appropriate column. Only those columns shall be used which are pertinent to the particular chart being constructed. Other columns may be added as required.

c. Item to be inspected column. The items listed in this column shall be divided into groups indicating the portion of the equipment of which they are a part, e.g., ''front'', ''left side'', ''engine'', ''turret'', ''video decoder'', auxiliary equipment.'' Under these groupings, the item to be inspected shall be identified by as few words, usually the common name, as will clearly identify the item, e.g., ''bumper'', ''gas can and mounting bracket'', ''front axle'', ''night vision scope.'' This column may be combined with the procedures column.

d. Procedures column. This column shall contain, not reference (references must be approved by the procuring activity), a brief description of the procedure by which the check is performed. Illustrations should be integrated with the tables whenever practicable. It shall contain all the information required to accomplish the checks and services including appropriate tolerances, adjustment limits, and instrument and gauge readings.

e. Equipment is not ready/available if: column. This column shall contain the criteria that will cause the equipment to be classified as not ready/available because of inability to perform its primary combat mission. An entry in this column will:

(1) Identify conditions that make the equipment not ready/available for readiness reporting purposes.

(2) Deny use of the equipment until corrective maintenance has been performed.

Figure 12-7 Definition of PMCS Columnar Entries

- Assembly and preparation for use
- Initial adjustments, daily checks, and self-test
- Operating procedures
- Operating auxiliary equipment
- Preparing for movement
- Operating instructions shown on decals and plates

Figures 12-8, 12-9, taken from MIL-M-38784B, and 12-10, taken from MIL-HDBK-63038-A (TM), show acceptable examples of these instructions.

1. Assembly and Preparation for Use

Normal assembly and preparation for use are covered in Chapter 11. If there are any special assembly tasks other than those discussed, you would include them in this paragraph.

2. Initial Adjustments, Daily Checks and Self-Test

Adjustment instructions cover the transition process from assembly and preparation for use to operation. This para-

MIL-M-38784B

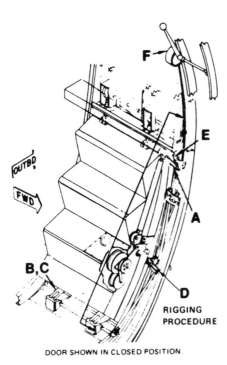

DOOR SHOWN IN CLOSED POSITION.

STEP 1

SUPPORT THE CREW DOOR AND REMOVE THE PIN (2) SECURING THE UPPER END OF THE TELESCOPING ARM (1) TO THE HOUSING (3).

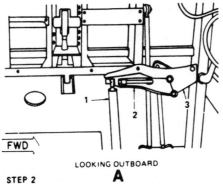

LOOKING OUTBOARD

A

STEP 2

PLACE THE DOOR IN THE JETTISON POSITION AND REMOVE THE DOOR

STEP 3

ROTATE THE LOWER TORQUE TUBE (4) TO THE JETTISON POSITION. CHECK THAT THE YOKES (5) AND BUSHING (6) ARE POSITIONED TO LET THE HINGE PINS (7) FALL FREELY.

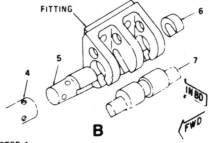

B

STEP 4

ADJUST THE STOP BOLT (8) AGAINST THE TUBE BOLT (9) WITH THE TUBE (4) HELD IN THE JETTISON POSITION

STEP 5

LOOKING FORWARD, ROTATE THE LOWER TORQUE TUBE (4) COUNTERCLOCKWISE 90 DEGREES TO THE DOOR KEPT POSITION

STEP 6

CLAMP THE TUBE IN THIS POSITION, AND ADJUST THE OTHER STOP BOLT (11) AGAINST THE TUBE BOLT (10) WITH TUBE (4) HELD IN DOOR-KEPT POSITION

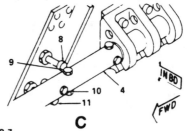

C

STEP 7

AT THE OUTSIDE OPERATING HAND, INSERT A RIG PIN THROUGH THE RIGGING HOLE IN THE BRACKET (12) AND THE LOWER HOLE IN THE LEVER (13)

STEP 8

CLAMP LEVER (14) IN POSITION SO THE GAP BETWEEN THE BOLT HEAD (15) AND LEVER (14) IS 0 00 TO 0 10 INCH.

STEP 9

MARK POSITION REFERENCE POINTS ON HANDLE AND ADJACENT STRUCTURE TO PERMIT CHECKING HANDLE POSITION THROUGH SUBSEQUENT RIGGING STEPS

Figure 12-8   An Example of an Operation Procedure Format

MIL-M-38784B

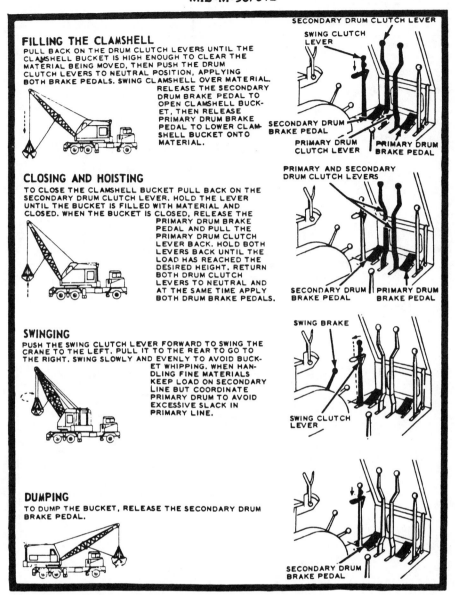

FILLING THE CLAMSHELL
PULL BACK ON THE DRUM CLUTCH LEVERS UNTIL THE CLAMSHELL BUCKET IS HIGH ENOUGH TO CLEAR THE MATERIAL BEING MOVED, THEN PUSH THE DRUM CLUTCH LEVERS TO NEUTRAL POSITION, APPLYING BOTH BRAKE PEDALS. SWING CLAMSHELL OVER MATERIAL. RELEASE THE SECONDARY DRUM BRAKE PEDAL TO OPEN CLAMSHELL BUCKET, THEN RELEASE PRIMARY DRUM BRAKE PEDAL TO LOWER CLAMSHELL BUCKET ONTO MATERIAL.

SECONDARY DRUM CLUTCH LEVER
SWING CLUTCH LEVER
SECONDARY DRUM BRAKE PEDAL
PRIMARY DRUM CLUTCH LEVER
PRIMARY DRUM BRAKE PEDAL

CLOSING AND HOISTING
TO CLOSE THE CLAMSHELL BUCKET PULL BACK ON THE SECONDARY DRUM CLUTCH LEVER. HOLD THE LEVER UNTIL THE BUCKET IS FILLED WITH MATERIAL AND CLOSED. WHEN THE BUCKET IS CLOSED, RELEASE THE PRIMARY DRUM BRAKE PEDAL AND PULL THE PRIMARY DRUM CLUTCH LEVER BACK. HOLD BOTH LEVERS BACK UNTIL THE LOAD HAS REACHED THE DESIRED HEIGHT. RETURN BOTH DRUM CLUTCH LEVERS TO NEUTRAL AND AT THE SAME TIME APPLY BOTH DRUM BRAKE PEDALS.

PRIMARY AND SECONDARY DRUM CLUTCH LEVERS
SECONDARY DRUM BRAKE PEDAL
PRIMARY DRUM BRAKE PEDAL

SWINGING
PUSH THE SWING CLUTCH LEVER FORWARD TO SWING THE CRANE TO THE LEFT. PULL IT TO THE REAR TO GO TO THE RIGHT. SWING SLOWLY AND EVENLY TO AVOID BUCKET WHIPPING. WHEN HANDLING FINE MATERIALS KEEP LOAD ON SECONDARY LINE BUT COORDINATE PRIMARY DRUM TO AVOID EXCESSIVE SLACK IN PRIMARY LINE.

SWING BRAKE
SWING CLUTCH LEVER

DUMPING
TO DUMP THE BUCKET, RELEASE THE SECONDARY DRUM BRAKE PEDAL.

SECONDARY DRUM BRAKE PEDAL

Figure 12-9 *Second Example of an Operation Procedure Format*

graph tells operators clearly where to position each control before applying power and tells them any pre-use inspections or actions needed. You explain how to switch the system ON, how to initiate a self-test, and how to set controls for the desired mode of operation.

3. Operating Procedures

This paragraph covers the routine, day-by-day use of the system. You tell the operators how to select each mode of operation and how to set controls to handle misfires or other emergency conditions. You also tell them how to remove the units from operation and how to switch the system to a STANDBY mode.

4. Operating Auxiliary Equipment

Many systems include auxiliary equipment such as motor/generators, air-pumps, cooling systems, or environmental control units (air-conditioners). This paragraph tells the operators how to use each auxiliary unit. You tell them only what they need to know to identify, connect, operate, and protect the auxiliary equipment and to protect themselves. If the auxiliary unit has its own manual make reference to it in this paragraph.

5. Preparing for Movement

Some systems are designed to be mobile or transportable. Others are fixed and are not designed for movement. If the

MIL·HDBK·63038·1A (TM)

MAINTENANCE OF RECEIVER ASSEMBLY

2-27.   FUNCTIONAL TEST OF DRIVE MOTOR

INITIAL SETUP

Tools   Equipment Conditions:
                                                   Reference
Power Supply 10-32 vdc, 0-50 amps   Drive motor removed,
    Part No. TC-32-50   TM 9-1005-200-20&P

1. SETUP POWER SUPPLY ON BENCH.

    a. Position power supply (1)
       on work surface. Connect
       power lead to 220 volt
       source.

    b. Position ON/OFF switch (2)
       to OFF.

    c. Rotate OUTPUT knob (3) to
       zero.

2. CONNECT INTERCONNECT HARNESS
   TO DRIVE MOTOR.

    a. Position interconnect
       harness (4) and drive motor
       (5) on work surface.

    b. Connect interconnect
       harness plug (6) to drive
       motor receptacle (7).

3. CONNECT INTERCONNECT HARNESS
   (4) TO POWER SUPPLY (1).

    a. Connect both black wires (8)
       to the negative post (9)
       on the power supply (1).

    b. Connect both red wires (10)
       to the positive post (11)
       on the power supply (1).

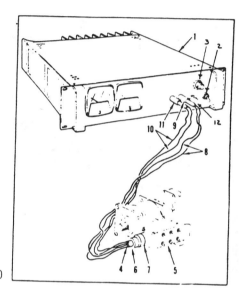

Figure 12-10   *Third Example of an Operation Procedure Format*

system must be prepared for movement after replacement, this paragraph tells the operators how to make the units ready for transportation. You need to give clear, step-by-step instructions and include illustrations. You show the operators how to make the system ready for movement and then how to get it ready to use after movement.

6. Operating Instructions Shown on Decals and Plates

Many times, the equipment you are describing in the manual includes some instructions on decals or plates fastened to the item. In this paragraph you include illustrations of each instruction decal and each instruction plate in the manual, making sure the illustrations are clear and readable when showing the location of the decal or plate. You explain

the use of these instructions and use the same instructions on the plate or decal that you use in the manual.

Section IV   Operation Under Unusual Conditions

The Operation Under Unusual Conditions section provides the operators with step-by-step instructions for operation under unusual and unexpected conditions. Equipment may be required to operate in less than desirable conditions. The following is a list of some unusual conditions:

● failed parts: logistic delay and urgent mission, etc.

● severe weather: sand, dust, smoke, rain, ice, mud, etc.

- off-road operation: fording streams, rocky terrain, etc.
- interference: radio-frequency interference, explosions, etc.

Emergency or unusual conditions may be artificial or natural; in either case, the operator needs to know how to ''get the job done'' under adverse and unusual conditions. Since the conditions are unusual, the operator does not have time to practice, and there is rarely time in training for many of these conditions. Because operators are under stress when they need these instructions, you need to write those instructions as simple as possible so that they are not misinterpreted.

Writing Style

When writing the Operation Chapter, you need to describe tasks in clear text and with key illustrations. The illustrations show the operator where the controls and indicators are. You need to write the elements of each task in simple active voice sentences. Use unambiguous action verbs to start each sentence and link the task elements to the illustrations. You should be able to get clear useful tasks from the LSA data base developed during equipment design.

Conclusion

The Operation Chapter tells the readers only what they need to know to use the equipment safely and properly. The chapter does not provide instructions pertaining to mission operating procedures. It does discuss operation of the equipment under both the usual expected conditions and unusual or unexpected conditions. Where needed, the chapter refers to associated equipment manuals or to other chapters of the manual.

The Theory of Operation Chapter in the Manual

Introduction

The Theory of Operation Chapter explains the theory of operation (how the equipment works) in a series of increasingly complex descriptions (top-down order). It starts with a functional diagram and discussion of the top level functions and the major items of the system. The top level diagram must identify all major functions and each major element of the system. The depth of theory discussion should be consistent with the maintenance concept and the type of manual (operator, field maintenance, depot maintenance, etc.). There is no need to explain detailed circuit theory if the user is not allowed to repair the item. The manual needs to tell the readers only what they need to know in order to perform their tasks.

Sections in Chapter

The usual section format for this chapter is as follows:

Section I	Introduction
Section II	General Description
Section III	Detailed Description

You use the pyramid structure to plan this theory description and to show the users the most important, top level theory. You then expand from this overall view to more general detail, and finally the specific details—thus, the pyramid structure.

Build on the foundations you lay in the earlier paragraphs to take the users from the known to the unknown, from the familiar to the unfamiliar. In the introductory paragraphs you use a simple top level block diagram and add more detail in each subsequent paragraph. The following is an example, using a car as the "system/hardware/equipment."

Top Level	Introduction	Blocks include Body, Electrical, and Motive subsystems
Intermediate Level	General Description	Blocks of Motive system include Motor, Transmission, Drive Shaft, Brakes, and Wheel components
Lowest Level	Detailed Description	Blocks of Wheel component to include Wheels, Tires, Bearings, Valve Stem Assembly, and Hub Caps

Section I   Introduction

The Introduction section gives the users an overview by telling them what the system does, what the major components are, and what the major functions are. This section explains how the equipment works so that the users and operators make the best use of it. The users, operators, and maintenance personnel need to understand the theory for their own safety and the safety of the equipment. Use the highest level functional block diagram to explain the top level theory, and then expand the detail of both illustrations and explanation in later sections of the manual. Figures 13-1, 13-2, and 13-3 taken from MIL-M-63038B (TM) show typical examples of top level descriptions. The words and pictures are planned to get the most communication possible from the page. This set of descriptions is often called the "simplified block diagram," and is part of the system description from which you can build a firm base and from which you can lead the users/operators.

Section II   General Description

The General Description section provides a functional block diagram (see Figures 17-2 and 17-3) which gives key details of each of the functional blocks shown in the introductory top level diagram and explains each block or function shown. This section then is the intermediate part of this chapter so you need to build on what you told the users in the introduction and not get bogged down into too many details.

You may use simplified schematics and logic diagrams in the general description section if, and only if, you go into full depth in the detailed description section. You also must remember that users must be familiar with the terms and symbols used in this section of the manual.

Section III   Detailed Description

The Detailed Description section includes the specific details, but only those details the users NEED TO KNOW! For example, if they have taken basic electronics courses, they know what a "flip-flop" does so you do not try to explain the deep theory of the "bi-stable multivibrator." You just say, "U-102 is the flip-flop which stores the last position of the XYZ Switch." This section includes common building blocks in electronics, hydraulics, and many technologies. But you do not try to re-explain the common building block theory, only just what the block does in the

2-2. 7.62-MM MACHINE GUN, M73

Mounts inside turret parallel with 90-mm gun assembly. Is lightweight, air cooled, metallic link belt fed, short receiver configuration. Recoils with gas assist boost. Is designed with quick-change barrel. Fires from retracted position.

Ⓐ JACKET ASSEMBLY WITH BEARING GROUP. Secures to front portion receiver assembly (trunnion block) and is designed for quick removal.

Ⓑ COVER ASSEMBLY. Feeds belt and positions and holds cartridges in position for chambering.

Ⓒ FEED TRAY GROUP. Serves as guide for belt to assist in positioning cartridges and provides directional control for link ejection.

Ⓓ BACK PLATE ASSEMBLY, HELICAL (DRIVING) SPRINGS AND GUIDE ROD GROUP. Absorbs recoil shock and provides energy to feed strip, chamber, and fire following round. Houses trigger sear, solenoid, and trigger safety. Acts as positive sear block for manual operation by solenoid and when weapon is operated by solenoid.

Ⓔ BARREL EXTENSION GROUP. Chambers and fires cartridge, locks and unlocks breech, and extracts spent cartridge case.

Ⓕ CHARGE GROUP. Charges weapon before loading, loading first round, and to recharge weapon in case of malfunction or stoppage. Assemblies on right or left side of receiver by repositioning slide connector, and buffer pivot pin.

Ⓖ RECEIVER ASSEMBLY. Serves as support for all major assemblies and groups. Houses action of weapon and through a series of camways, controls functioning of barrel extension assembly, breechlock assembly, and buffer assembly.

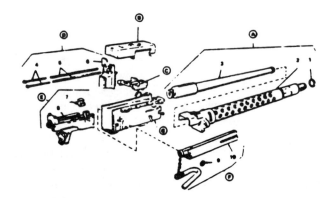

Ⓐ JACKET ASSEMBLY WITH BEARING GROUP
  1. BARREL BEARING LOCK
  2. JACKET ASSEMBLY WITH BEARING
  3. BARREL ASSEMBLY

Ⓑ COVER ASSEMBLY

Ⓒ FEED TRAY GROUP

Ⓓ BACK PLATE ASSEMBLY, HELICAL SPRING AND GUIDE ROD ASSEMBLY
  4. GUIDE ROD ASSEMBLY
  5. COMPRESSION HELICAL SPRING (DRIVING SPRING)
  6. BACK PLATE ASSEMBLY WITH SOLENOID

Ⓔ BARREL EXTENSION GROUP
  7. BREECHBLOCK ASSEMBLY
  8. BARREL EXTENSION ASSEMBLY

Ⓕ CHARGER GROUP
  9. RETAINING RING
  10. CHARGER ASSEMBLY

Ⓖ RECEIVER ASSEMBLY

2-4

Figure 13-1 A Typical Mechanical System Description

2-16. RECEIVING FUNCTION

Receives reflected rf energy from target. Channels energy to processing circuits for display on radar screen.

(A) RADAR ANTENNA. Parabolic reflector which collects reflected energy from targets along antenna boresight axis. Mechanically positioned to search for and track targets. Modulates signal to develop azimuth and elevation pointing-error signals for positioning.

(B) WAVEGUIDE COMPONENTS. Permit use of a single antenna for both transmitting and receiving. Circulate energy from transmitting function to antenna and from antenna to tr tube. Circulate any energy reflected from tr tube to resistive load.

(C) TR TUBE. Short-circuits rf pulse leakage to prevent its entering receiving section when rf leakage may be of sufficient amplitude to damage paramp or image rejection mixer diodes. Keep-alive voltage maintains ionization during transmission but allows received weak signals to pass through tube without ionization. Voltage applied only during radar on operation.

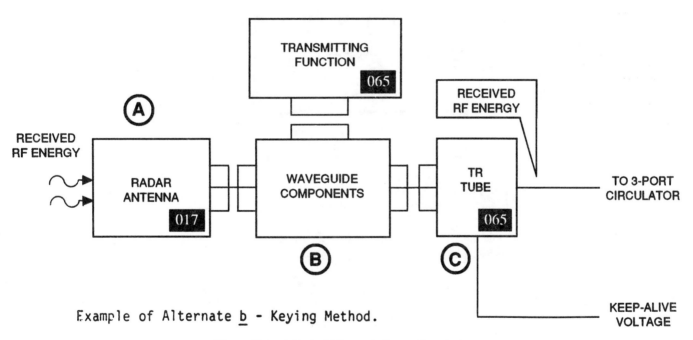

Example of Alternate b - Keying Method.

Figure 13-2 *A Typical Electronic System Description*

system and how it connects to and works with the other blocks. This is the part of the Theory Chapter where logic diagrams and schematics are most useful, but you must make sure you tell users what they need to know in terms they understand. You amplify the theory to the level the users need to know to provide a firm understanding of what the equipment does, and to avoid trauma to themselves or damage to the system.

Writing Style

When writing the Theory of Operation Chapter (Introduction, General, and Detailed), you start with a clear, simple sentence such as:

The "whatzit block" accepts primary in power from the isolation transformer and converts it to primary out power which is distributed to the XYZ and the ABC.

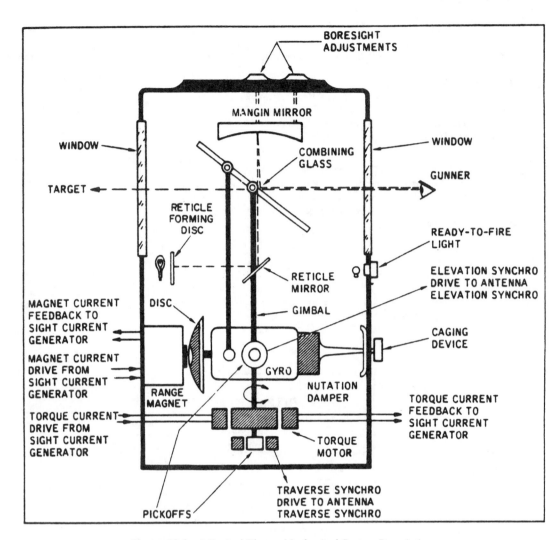

Figure 13-3 *A Typical Electro-Mechanical System Description*

This clear topic sentence should show the key relationships, interconnections, and functions of each part of the system. The rest of the sentences in each paragraph build on that topic sentence. You need to plan a consistent order for each theory description and keep that consistent order in the Introduction, General and Detailed Description sections. Of the two patterns—spatial (outside to inside, top to bottom or right to left) and cause/effect (starting with the first function and showing its consequences)—the spatial-ordered pattern is usually the better choice for the Theory of Operation Chapter.

Conclusion

In the Theory of Operation Chapter you explain how the equipment works. For clarity you start with the "big picture" and keep adding detail. You would use a level of detail and technology you know the targeted readers understand and need to perform their functions. Thus you carry the clear simple writing style from the introduction, through the general description of the theory into the detailed description of the theory of operation.

Chapter 14

The Maintenance Chapter in the Manual

Introduction

The Maintenance Chapter tells users/operators and maintenance personnel what they need to know to keep the system working and to repair the system when it breaks down. You tell users clearly and accurately how to keep the system in good running order and how to restore it to proper operation.

Sections in the Chapter

The Maintenance Chapter usually includes three sections:

Section I   Lubrication Instructions
Section II   Troubleshooting Procedures
Section III   Maintenance Procedures

If auxiliary equipment is used with the system, you may have to add a maintenance chapter for those items.

Service, Corrective Maintenance, and Preventive Maintenance

Maintenance is that class of activities which include tasks to fix a broken system (called corrective maintenance or CM) or keep a system from breaking (called preventive maintenance or PM). PM is always scheduled and planned but CM is never scheduled but should be carefully planned. Three sub-classes may be used: Service (preventive tasks such as lubrication, replenishing supplies, etc.), PM, and CM. Of the three sections, Section I (Lubrication Instructions) and Section III (Maintenance Procedures) describe PM. Section II (Troubleshooting Procedures) and Section III describe CM. Each maintenance instruction tells users where, what, and how as shown below:

- WHERE: to find the indicators and controls to set a control, to apply the lubricant, to connect the test equipment, etc.
- WHAT: tool to use, meter to read, oil to apply, reading to expect, and what constitutes an error, etc.
- HOW: to do the task, to find the adjustment point, to disconnect the assembly, to set the voltage, etc.

You must read the customer's specification carefully because you may need to include an alphabetical index at the beginning of each section. You give both operators and maintenance personnel locating instructions and tell them what they will do, what they need to do, and what to use to do it with. The instructions clearly describe each step so the operators can easily learn to do the task. If you listed Preventive Maintenance Checks and Services (PMCS) in the Operation Chapter, you must now tell operators how to do the PMCS as well as how to repair the system. However, the procedures you provide must limit the level of detail to what the operators know. The maintenance concept (expressed, for example, in the maintenance allocation chart, or MAC) should define what tools and test equipment the operator is capable of using.

Section I   Lubrication Instructions

Lubrication is one of the task types usually assigned to the sub-class called Service. The customer may require separate lubrication orders for each task or allow a table to list each task. Regardless, in the Lubrication Instructions section you write the service and lubrication instructions clearly and state whether tasks are mandatory or optional or ''on condition'' (on condition tasks are done if an inspection shows the task is needed). If no lubrication is needed, you must say that. Figure 14-1 taken from MIL-HDBK-63038-1 shows typical lubrication instructions.

Section II   Troubleshooting Procedures

The Troubleshooting Procedures section tells the operators and maintenance personnel clearly how to identify symptoms of a system failure and tells them how to isolate the problem to identify the faulty replaceable unit. You also need to tell them how to fix the system quickly, safely, effectively, and properly. The step-by-step process is usually called ''troubleshooting.'' When writing troubleshooting instructions, base your instructions on the symptoms you reasonably expect the operator or maintenance personnel to see, or hear, or feel, or smell. For easily discernable symptoms give only the tasks the operator or maintenance personnel (the expected reader of this level manual) is allowed to do. List tests and inspection in logical order for each malfunction covered. Usually the most logical sequence is the order in which the faults are most likely to be sensed.

Include references to the service or PM sections as needed to complete repairs described. Plan the troubleshooting by considering the location of assemblies, sub-assemblies, controls, indicators, and replaceable units. You show each

MIL-HDBK-63038-1

c. Use facing pages.

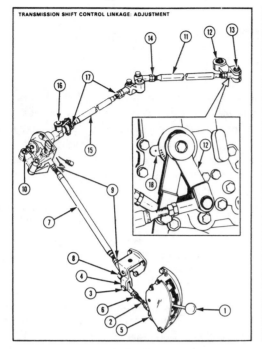

d. Combine information in the location, item, indicator, and locator illustrating columns if it can be done without confusing the reader.

LUBE IT !

DON'T FORGET; KEEP THE WEA-
PON CLEAN AND LUBED, EVEN
WHEN IT WILL BE UNUSED FOR
A PERIOD OF TIME.

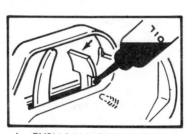

4. TURN LAUNCHER UPSIDE
DOWN AND LUBE THE
SAFETY DETENT. IT'S IN
THE RECEIVER IN FRONT
OF THE SAFETY.

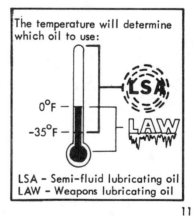

The temperature will determine
which oil to use:

0°F

-35°F

LSA

LAW

LSA – Semi-fluid lubricating oil
LAW – Weapons lubricating oil

11

Figure 14-1 *A Typical Set of Lubrication Instructions*

malfunction and the series of steps it takes to locate each malfunction. Figure 14-2, taken from MIL-HDBK-63038-1, shows a typical troubleshooting table.

This table uses an indented format rather than the matrix or column format. The headings (Operation, Normal Indication, and Corrective Procedure) show the indenture

level. MIL-HDBK-63038-1 includes other table formats and techniques which the contract may require instead of the example shown, so you need to know the contract requirements. In most cases, you may have to precede troubleshooting procedures with introductory statements such as the following:

d. *Troubleshooting procedures.* A troubleshooting procedure consists of action, normal indication, and corrective action statements.

Example of Troubleshooting Procedure

Step	Operation Normal indication Corrective procedure		
	NOTE The key numbers shown below in parentheses refer to figure 5-1 unless otherwise indicated.		
1.	Quick-Disconnect Couplings and Handpump Air Purge.		
	Open the handpump access panel (10, fig. 1-3) and insert the handpump handle into the handpump.		
	Handpump unloading valve (16, fig. 2-11)	full cw.	
	HAND PUMP PRESSURE UNLOADING VALVE (13)	partial cw.	
	Actuate the handpump handle six to seven strokes to purge the handpump of air.		
	Connect hose assembly MS28762-4-0250 with quick-disconnect coupling 9194683 to the HAND PUMP PRESSURE PORT (12).		
	Actuate the handpump handle to purge the handpump system of air.		
	Hydraulic fluid flows from the open-end hose assembly.		
	HAND PUMP PRESSURE PORT, quick-disconnect coupling.		
	Remove the hose assembly.		
2.	Handpump Low Pressure.		
a.	HAND PUMP PRESSURE UNLOADING VALVE	full cw.	
	Actuate the handpump handle until the 0–100 PSI GAGE (1) indicates 10 psig; maintain this pressure for 3 minutes.		
	The 0–100 PSI GAGE reads 10 psig. No fluid leakage is observed.		
	Gage, handpump, HAND PUMP PRESSURE UNLOADING VALVE.		
b.	Actuate the handpump handle until the 0–100 PSI GAGE indicates 90 psig, and maintain this pressure for 3 minutes.		
	CAUTION Do not exceed the 100-psig indication on the 0–100 PSI GAGE if the gage protector is not functioning properly.		
	The 0–100 PSI GAGE reads not less than 88 psig. No fluid leakage is observed.		
	Gage, handpump, HAND PUMP PRESSURE UNLOADING VALVE.		
c.	Actuate the handpump handle until the 0–1000 PSI GAGE (2) indicates 500 psig. Observe all connections for leakage.		

Figure 14-2 A Typical Table of Troubleshooting Instructions

The table lists the common malfunctions which you may find during the operation or maintenance of the (insert type of equipment or system) or its components. You should perform the tests/inspections and corrective actions in the order listed.

This manual cannot list all malfunctions that may occur, nor all tests or inspections and corrective actions. If a malfunction is not listed or is not corrected by listed corrective actions, notify your supervisors.

The contract may also require a "Symptom Index" preceding the troubleshooting guide. Figure 14-3, taken from MIL-HDBK-63038-1, shows a typical symptom index which leads the operator from the discernable symptom to the proper starting point in the troubleshooting process.

When you create your troubleshooting chart, you need to consider the following:

MIL-HDBK-63038-1

STEP	INSTRUCTION	INDICATION	YES	NO	REMARKS
	WARNING In the following procedure, operation of the armament bay doors and missile launchers is a potential safety hazard to personnel in the vicinity of the aircraft. Adhere to all safety precautions iaw T.O. 1F-106A-2-12.				
1	[TM 27-2] ATS Check: Perform	[Master warning panel] MISFIRE: Goes off when TEST MISFIRE FWD and TEST MISFIRE AFT are set to down (off)	2	[054] [Launcher harness] [596] [Intervalometer] [Misfire relay]	MISFIRE remains on after the fire signal if either a ground path is maintained to the misfire relay hold-in coil or +28 v is maintained to the misfire relay pull-
	NOTE Perform steps 2 and 3 on suspected faulty rail only, or all rails for a complete system check.				
2	[054] Missile gone switch: Actuate to the up position for entire step [Missile simulator] MISSILE SIMULATE: GONE [27-2] ATS Check: Perform with S96 TEST MISFIRE switches in the down (off) position	[Master warning panel] MISFIRE: Blinks on momentarily when the fire signal is delivered	3	[054] [Launcher harness] --------	Missile gone switch circuit in launcher may be faulty
3	[054] Missile gone switch: Actuate to the up position for entire step [Missile simulator] MISSILE SIMULATE: PRESENT [TM 27-2] ATS Check: Perform	[Master warning panel] MISFIRE: Remains on after the fire signal is delivered [Armament bay] Doors: Remain open Launchers: Remain extended	▲	[054] [Launcher harness] -- -- -----	Checks the ability of the system to stop the armament sequence when a misfire condition is simulated

Figure 14-3 A Typical Symptom Index to Troubleshooting Instructions

- The maintenance concept developed for the equipment
- The ease and simplicity of testing
- Elemental reliability
- Test time
- Type of test: automatic, manual, a little of each
- Available test equipment
- The user's environment
- The repair technician's experience
- Limitations on system turnaround

The three phases of the troubleshooting process are:

1. Fault Discovery: "It's broken!"
2. Troubleshooting Entry: "Start finding it!"
3. Troubleshooting   "Find and fix it!"
   Accomplishment:

Each phase requires a slightly different method of writing, and Figure 14-4 taken from MIL-HDBK-63038-1A (TM) compares the phases and the methods. The troubleshooting section must help the user/operator rapidly find and fix the problem. Each troubleshooting procedure has 10 desirable characteristics:

1. Failure verified as first step in troubleshooting process
2. Technician not left high and dry by troubleshooting data
   a. Sufficient descriptive data provided so technician knows what to do and why
   b. When troubleshooting procedures are not successful, references provided to troubleshooting diagrams or functionals so troubleshooting of rare or difficult problems can be continued easily
3. Troubleshooting data keyed to failure symptoms encountered during operation and maintenance procedure performance
4. All data complete and correct
5. Data easy to update to account for new equipment or newly discovered problems
6. Specific troubleshooting information easy to locate quickly
7. Data and format simple to use and easy to understand
8. Only data required to understand/perform specific troubleshooting task included
9. Parts verified faulty before replacement—whenever possible
10. Proper operation verified after parts replacement to ensure trouble has been repaired

MIL-HDBK-63038-1A (TM)

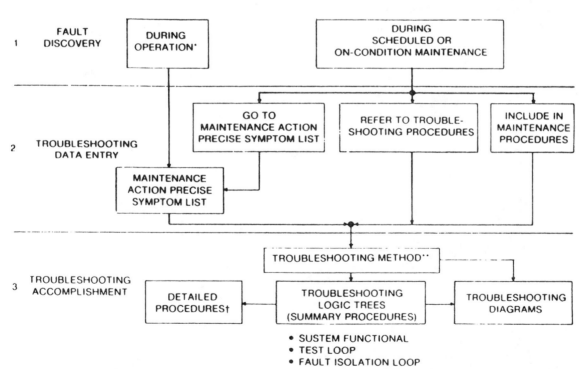

Figure 14-4 *Different Writing Methods for Troubleshooting*

Good troubleshooting procedures start with general, easily discerned symptoms and then systematically guide the user to the failed, replaceable unit. The three phases of the troubleshooting process are described in the following paragraphs.

1. Fault Discovery

Operators and maintenance personnel usually discover faults in one of four ways: three during operation and one during maintenance. During operation, the fault may trigger an alarm, may cause a clearly sensed symptom, or may cause another operator to detect a failure in the equipment. During maintenance, the operator or maintenance person may find that a parameter measures out-of-tolerance or a condition requires correction. The failure, then, has caused the out-of-tolerance condition, the leak, or some other symptom found through testing or inspecting. In this paragraph you need to be specific and thorough when describing symptoms to the operators and maintenance personnel.

2. Troubleshooting Entry

After explaining how to discover that a fault exists, this phase tells operators how to start locating the problem. You identify and list a set of symptoms that are complete enough to lead the maintenance personnel to the fault with as little ambiguity as possible. The Maintenance Action Precise Symptom (MAPS) list is one way to show the symptoms

to the operators. Figure 14-5, taken from MIL-HDBK-63038-1, shows a typical MAPS list for a simple system and Figure 14-6, taken from MIL-HDBK-63038-1A (TM), shows a typical MAPS list for a more complex system. The MAPS list maybe tabular or branching, but whichever form you use, each MAPS list must:

- Contain fault symptoms for each known malfunction.
- Catalog symptoms by method of detection, if that helps make the list more useful.
- List symptoms by sub-system or by other clear elements when necessary for more complex systems.

You incorporate a troubleshooting entry from maintenance by:

- Integrating the entry into maintenance instructions ("on condition that . . . , do . . . ").
- Referring to separate troubleshooting steps directly.
- Referring by symptom to the MAPS list.

An integrated "on condition" entry is preferred, because in this way you do not make operators turn to another part of the manual to get the information. You need to use that method whenever possible, but if the fault is complex or unusual, you should refer to other troubleshooting entry guides. Reference to MAPS is the least desirable method, because just referring to MAPS lists forces the operators to carry out another search process before doing the actual troubleshooting.

SYMPTOM INDEX   **MIL-HDBK-63038-1**

	Troubleshooting Procedure (Para)
COOLING SYSTEM	
Radiator	
Boils over	3-12
Leaks	3-17
Temperature Gage	
No indication	3-29
Runs cold	3-11
Runs hot	3-13
ENGINE	
Misses	3-34
Overheats	3-72
Won't start	3-27
EXHAUST SYSTEM	
Excessive smoke	3-33
Water vapor	3-33

Figure 14-5 *Typical MAPS List for a Simple System*

MIL·HDBK·63038·1A (TM)

CODE	DESCRIPTION	FAULT ISOLATION PROCEDURE
DM-1	080 unit target marker circle range incorrect - radar dominant	3-12
DM-2	080 unit target marker circle elevation incorrect - IR dominant	3-12
DN	□ TSD target bug range and HSI target range incorrect - data link	8-17
DN-1	□ TSD target bug bearing and HSI target bearing incorrect - data link	9-3
DN-2	□ HSI miles indicator incorrect - data link	8-17
DT-5	□ AMI command Mach incorrect - manual	8-10
DIGITAL COMPUTER		
EA	Computer malfunction - can be cleared	6-3
EA-1	Computer ASIA information incorrect	2-10
EA-2	Computer value of WSEM parameter/CCM signals incorrect	2-4
EA-3	Temperature call up incorrect on AVVI or 980 unit	. . .
EB	Computer malfunction - can not be cleared	6-4
EB-1	Computer manual steering erroneous - all modes	3-16
EB-2	Computer manual and AFCS steering erroneous - all modes	8-5
ARTIFICIAL HORIZON		
FA	Artificial horizon - no erection - roll	8-2
FA-1	Artificial horizon - no erection - pitch	8-2
FC	Artificial horizon - drifts/tumbles - pitch	8-1
FD	Artificial horizon - incorrect erection - pitch	8-2
FJ	Artificial horizon - missing	3-12
FJ-1	Artificial horizon - distorted	3-12
FK	905 unit annunciator incorrect	8-4
RADAR SEARCH AND DETECTION		
GA	B-sweep missing - on and stby	3-1
GA-1	B-sweep missing - on	3-1
GA-2	B-sweep missing - stby	3-1
GB	B-sweep intensity incorrect	3-2
GC	B-sweep super search - inoperative	3-3
GC-1	B-sweep super search width incorrect	3-3

Figure 14-6   Typical MAPS List for a More Complex System

3. Troubleshooting Accomplishment

This phase describes troubleshooting methods by use of tables (branching or narrative), logic trees, or troubleshooting diagrams. The Logic Tree is a programmed instruction to the operators because it gives them an instruction, asks a question about the results of taking an action and makes sure that the only answer to the question is either a YES or a NO. You then refer operators to the next step, depending on the answer.

Logic Trees begin with clear symptoms and flow through the most probable faults and the most probable causes. Rare or improbable faults are excluded. You refer operators to other troubleshooting aids such as diagrams or sequential lists of "replace/test/replace" instructions. This procedure is repeated until the problem is fixed.

Each troubleshooting process must have the most effective mix of the following characteristics:

- Test Time (as little as possible)
- Test Access Time (as little as possible)

- Test Equipment (as simple as possible)
- Reliability of Replacement (as high as possible)
- Validity of Test (as high as possible)

Most troubleshooting processes use some form of "half-split" logic. That is, each test should eliminate about half of the possible failures which could cause the symptoms. Figure 14-7 is an example.

In this example, assume the output signal at G is missing and the indicators show that the input signal at A is correct. The operators could chase the good signal by measuring at B,C,D (and so on) until they find the bad signal. If the failed unit is the last block, that means five tests have to be performed. (The same logic holds if the operators chase

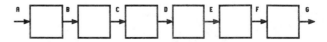

Figure 14-7   Using the "Half-Split" Troubleshooting Process

a bad signal back and find the first block bad.) Or the operators may "divide and conquer," by first measuring the signal at D. If it is good, they measure the signal at either B or C. Again, it only takes two or three tests to find the failed block. In all cases, the operator assumes there is only one fault because in the majority of cases only one fault exists and it is just "too hard" to describe the symptoms for multiple faults in the manual. Logic Trees must contain well-designed procedures and each procedure must tell the operators the location, the item or indicator, the action to take, the expected indication, and the action to take if the indication differs from the one expected.

Troubleshooting diagrams can supplement the Logic Tree or they can stand alone. The more common troubleshooting diagrams are the System Functional Diagram, the Test-loop Functional Diagram, and Fault Isolation Diagram.

System Functional Diagrams show the operation and functions of the system and divide the system into "real-world" functional blocks. Test-loop Functional Diagrams show all circuits included in the loop. They are oriented toward stimulus-response patterns. They include built-in or automatic test functions, and identify all test conditions (preset, setup, test stimuli, items tested, test equipment connections, expected results, and all other pertinent data).

Fault Isolation Diagrams are similar to Test-loop Diagrams but are symptom or indication oriented rather than stimulus-response oriented. Table 14-1 compares the criteria used to select the test method for troubleshooting procedures. If the customer does not specify the form, you must choose the best based on these criteria. MIL-HDBK-63038-1 describes one method for designing a troubleshooting Logic Tree:

1. Using schematic diagrams, functional diagrams, available engineering documentation such as test specifications, design descriptions, trouble and failure reports, field squawks, and any other available documents, study the system in detail. Understand the system operation and note what functions are performed by the units associated with the system. If possible, visit user sites and obtain actual information on fault

Table 14-1   *Troubleshooting Method Selection Criteria*

Selection Criteria	Troubleshooting Method	
	Logic Trees	Troubleshooting Diagrams
Hardware Complexity		
Simple		•
Complex	•	•
Quantity Produced		
Few		•
Many	•	•
Built-in Test		
Little	•	•
Extensive	•	
No. of Personnel		
Few		•
Many	•	•
Personnel Experience or Training		
Little	•	
Extensive		•
Tasks		
Simple		•
Complex	•	
Extent of Parts Replacement		
Unit Assembly Module	•	•
Component Part		•
Mission Importance		
Unimportant		•
Essential	•	
Frequency of Fault Occurrence		
Rare		•
Common	•	•
Environment		
Sheltered	•	•
Exposed	•	
Working Area		
Ample	•	•
Limited	•	

symptoms and causes resulting from hardware usage. Understand the user's method of troubleshooting and his or her problems.

2. Knowing all the variables about the system such as test equipment available, repair level, technician experience level, number of technicians, etc., perform a fault analysis on paper, listing all problem areas and faults that could possibly occur. Actual usage of the hardware during fault analysis is very helpful and should be used as much as possible. Because fault isolation must be accomplished in the least amount of time, the logic tree should begin with tests of components with the least reliability or tests which require the least time to be performed.

3. All components that produce the same malfunction symptom must be considered as possible causes of the fault. All components should be associated through their failure mode to a malfunction symptom. Once the failure mode of the component is determined, determine all outputs that would be incorrect for each failure and describe what the incorrect measurement would be.

4. Note any effect on the other outputs downstream from the failing output.

5. Develop the logic tree based on tests, measurements, and decisions that must be made in order to reach the final outcome of isolating the fault to a replaceable component related to the symptom. All of the most probable faults and causes should be considered.

6. Include any information that will aid the technician such as waveshapes, voltage levels, references to test diagrams, functional diagrams, text, etc., and alignment procedures, checkout procedures, or other scheduled maintenance procedures. The end result must be the repair of the system.

7. List any self-tests that are associated with the system and understand to what extent the self-test is conducted on the system. Self-test schemes should be the prime troubleshooting tool, with manual troubleshooting by logic tree taking over where self-test leaves off or fails to locate malfunction.

8. Prepare a rough draft of the logic tree and include safety precautions such as notes, cautions, and warnings.

9. Build the procedure using system self-test first before using external test equipment. Avoid as much as possible procedures that use excessive test equipment or require measurements to be made in locations difficult to access.

10. Ensure that actions required of the technician are reasonable to perform. For example, do not ask the technician to connect a meter at one point and set a switch at another point where he or she cannot observe the meter reading. Be practical.

11. When a rough draft procedure is completed, validate proper operation of the procedure by the most efficient and practical method—preferably on the actual hardware. Whenever possible, create the fault conditions on the system by inserting fault and verifying results and corrective actions.

12. Include all faults possible in the MAPS.

Maintenance personnel cannot repair the system if they do not know how to "open it up" and "take it apart" and "put it back together" in order to replace the failed part. You need to keep these access instructions simple and specific, using illustrations as needed. List all tools and parts needed and warn the maintenance personnel if there is a possibility of trauma or system damage. You may include these access steps in the troubleshooting/repair process or in separate paragraphs. Furthermore, you must properly reference them in the manual.

The final part of any troubleshooting process is the repair and verification test. You should show a detailed test that verifies that the repair has really fixed the system. This "go-path" test should give the maintenance personnel confidence that the system is working properly. This test may be the same one used "before operation" as a confidence test. If each step in the test causes the correct response, the maintenance person goes on. If any step causes a false response, he/she goes back to the troubleshooting process.

Section III   Maintenance Procedures

The Maintenance Procedures section contains detailed descriptions of each scheduled task and CM. This section may contain each of the following as needed:

- Introduction
- Inspection
- Checks, Adjustments, Alignments
- Removal and Installation
- Disassembly and Assembly
- Repair Procedures
- Clean and Test Procedures

Figure 14-8, taken from MIL-M-63038A (TM), shows a typical CM task description. This figure is only an example; you need to use your creative skills to show the users/operators/maintenance personnel clearly the WHEN, WHAT, and HOW of the system.

Writing Style

When writing the Maintenance Chapter, you write directions that are keyed to illustrations to tell the maintenance personnel clearly what to do. You get most of the information for the tasks, maintenance tools, support equipment and supplies from the Input Records D and D1 if LSAR data is available. You select clear and unambiguous verbs to start each task and choose illustrations that show the maintenance person where to check equipment and what tools to use. Because the customer may spend more to maintain a piece of equipment through its life cycle than to buy it originally, clear and easy-to-follow maintenance instructions are very important.

Conclusion

The Maintenance Chapter provides a detailed description of each maintenance task which the operator or approved maintenance person may perform at the assigned level.

MIL-M-63038A(TM)

REMOVING/INSTALLING TRACK SHOE

1 Decrease track tension.

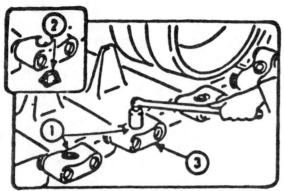 

2 Remove bolts (1) retaining wedges (2) on all four end connectors (3) holding track shoes.

3 Install end connector puller (4) and tighten screws (5) against ends of track link pins.

4 Move all four end connectors out approximately 1 inch, but do not remove.

 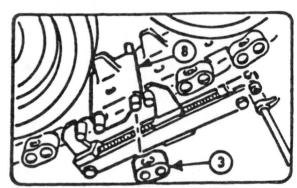

5 Install track fixtures (6), on both sides of shoe being removed, engaging track link pins (7) on adjoining shoes.

6 Remove end connectors (3) and track shoe (8).

7 Install new track shoe by reversing procedures 6, 5, 4, and 2.

8 Tighten wedge screws to 75-86 ft lbs if torque wrench is available.

Figure 14-8 *Typical CM Procedure (Replacement Task)*

Military manuals provide clear-cut allocation of tasks at each level or echelon of maintenance. The manual should discuss only those maintenance (unscheduled corrective and scheduled preventive) tasks allocated at the levels covered in the manual. For example, the owner's manual for a VCR will tell the owner how to clean the tape drive, but not how to align the record head or read head assemblies. The Maintenance Chapter includes discussion of appropriate lubrication, alignment, inspection, troubleshooting, and repair tasks.

Chapter 15

The Parts List Chapter in the Manual

Introduction

The Parts List Chapter tells users how to identify each item and each part needed to support the system at the covered (the intended user) maintenance level.

Sections in the Parts List

Various types of parts lists may include the following:

- Reference designators for each item (assembly, replaceable unit, electrical, electronic or mechanical part)
- Name of the item
- Brief description of the item
- Manufacturer's name or federal supply code for manufacturer (FSCM)
- Manufacturer's address
- True manufacturer's part number when another vendor's number is listed
- Quantity used in the system
- NATO stock number (NSN)

Levels in the Parts List

When writing a manual on a system/hardware/equipment, the parts list will help you tell users how the components, assemblies, units, items, and parts fit in relation to the system or the "end item." The levels, in order, include:

- Part (non-repairable item by design or by choice)
- Assembly (module, circuit card assembly, chassis, rack)
- Unit
- Equipment
- Component
- Sub-system
- Segment
- System

The end item described in the manual may be a single unit (chassis) or a multi-rack, multi-site system. The parts lists tell users what is where and what goes into what. In many cases, clear illustrations help to tell users what they need to know. These parts lists reinforce the descriptions covered in earlier chapters in the manual. The lists often give "ordering information" so users may replace failed items/parts. Some lists tell users what they need but do not have, and the users need to know what to order, if authorized by other documents such as a Table of Allowances (TA). Each

of these parts list types has a slight variation. The logisticians, engineers, and illustrators can assist you in preparing the Parts Lists Chapter.

Format for Parts Lists

Parts lists are called special names for specially defined content. Table 15-1 shows the most common military parts lists and their companion specifications.

The engineering parts list is the original source of information for most of these specialized parts lists. You need to understand the customer's requirements and the applicable specification designated in the contract.

Example of Parts Lists

One military specification covering preparation of operator's manuals (MIL-M-63036) calls for two appendices including four of these list types. Figures 15-1, 15-2, and 15-3 show examples of one such appendix including the Component of End Item List (COEIL) and Basic Issue Item List (BIIL). Figures 15-4 and 15-5 show another appendix containing an Additional Authorization Lists (AAL). Figures 15-6 and 15-7 show an appendix containing an Expendable Supplies (ES) and Materials List (ML).

Illustrating the Parts List

Some of the lists are not illustrated. Others use figures (line drawings, exploded views, cut-away views, selectively toned screen photographs, etc.) to communicate the information. Other lists, such as the COEIL and the BIIL, may show pictures of each component along with the formatted lists. You need to plan your lists to use the clearest, simplest illustration applicable.

Conclusion

When writing the Parts List Chapter for more complex systems, you will need specialized help from design engineers, provisioning specialists, technical illustrators, and customer logisticians. The lists you build will be the most common definition of parts published for the system/hardware/equipment. You use the equipment tree, the engineering parts lists, the support items lists, and vendor manuals to assemble the lists of items needed for each specified parts list for the manual. If you are required to illustrate your parts lists, you need to understand how to illustrate the lists "top down, break down" so that they contain all the line items needed.

Table 15-1 *The Most Common Type of Parts Lists*

Parts Lists Name	Acronym	Military Specification
Additional Authorized List Assembly Parts List	AAL	MIL-M-63036
Basic Issue Items List	BIIL	MIL-M-63036
Components of End Item List	COEIL	MIL-M-63036
Expendable Durable Supplies and Materials List	EDS&ML	MIL-M-63036
Manuals, Technical: Repair Parts and Special Tools List	RPSTL	MIL-STD-335
Illustrated Parts Breakdown	IPB	MIL-M-38807

APPENDIX C

COMPONENTS OF END ITEM
AND
BASIC ISSUE ITEMS LISTS

Section I.   INTRODUCTION

C-1.   SCOPE.

This appendix lists integral components of end item and basic issue items for the Antenna Group OE-361(V)1/G (Single Carrier Feed) and the Antenna Group OE-361(V)2/G (Multiple Carrier Feed) to help you inventory items required for safe and efficient operation.

C-2.   GENERAL.

This Components of End Item and Basic Issue Items Lists are divided into the following sections:

a. Section II. Integral Components of the End Item. These items, when assembled, comprise the Antenna Group PE-361(V)1/G or the Antenna Group OE-361(V)2/G and must accompany them whenever they are transferred or turned in. The illustrations will help you identify these items.

b. Section III. Basic Issue Items. These are the minimum essential items required to place the Antenna Group OE-361(V)1/G or the Antenna Group OE-361(V)2/G into operation, to operate them, and to perform emergency repairs. Although shipped separately packaged, they must accompany the Antenna Groups during operation and whenever they are transferred between accountable officers. The illustrations will assist you with hard-to-identify items. This manual is your authority to requisition replacement BII, based on TOE/MTOE authorization of the end item.

C-3.   EXPLANATION OF COLUMNS.

The following provides an explanation of columns found in the tabular listings:

a. Illustration. This column is divided as follows:

(1) Figure Number (Fig. No.). Indicates the figure number of the illustration on which the item is shown.

Figure 15-1 *An Introduction to an Appendix Using COEIL and BIIL*

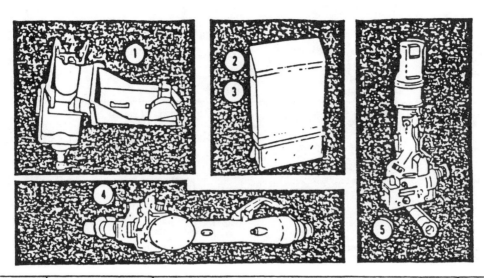

(1) Illus Number	(2) National Stock Number	(3) Description FSCM and Part Number	Usable On Code	(4) U/M	(5) Qty rqr
1	1005-00-706-8880	MOUNT, MACHINE GUN cal .50 (in mount on cupola) (19204) 7068880	PAA	EA	1
2	1240-00-344-4643	PERISCOPE: M27 (chief of section) (stowage box cab wall) (19200) 7633132	PAA	EA	1
3	1240-00-509-2743	PERISCOPE: M45 (driver's) (stowage box driver's compartment) (19200) 8213430	PAA	EA	3
4	1240-00-491-9676	TELESCOPE, ELBOW M118CA1 (in mount M146) (19200) 10559855	PAB	EA	1
		OR			
	1240-00-819-4520	M118C (19200) 10512985	PAC	EA	1
5	1240-00-864-2930	TELESCOPE, PANORAMIC: M117 (in mount M145 or telescope box) (19200) 7660400	PAA	EA	1

Figure 15-2 *An Example of a BIIL*

(1) ILLUSTRATION		(2) NATIONAL STOCK NUMBER	(3) DESCRIPTION		(4) LOCATION	(5) USABLE ON CODE	(6) QTY REQD	(7) QUANTITY	
(A) FIG NO.	(B) ITEM NO.		PART NUMBER	(FSCM)				RCVD	DATE
30		5995-01-182-2063	CABLE ASSEMBLY, RF (W115) A3020801-002	(80063)	STORED IN STORAGE BOX ASSEMBLY, A3020113-001		1		
31			CABLE ASSEMBLY, RF (W119) A3020801-001	(80063)	STORED IN STORAGE BOX ASSEMBLY, A3020113-001		1		
32		5995-01-182-2064	CABLE ASSEMBLY, RF (W122) A3020801-003	(80063)	STORED IN STORAGE BOX ASSEMBLY, A3020113-001)		1		
33			CABLE ASSEMBLY, RF (W123) A3020801-004	(80063)	STORED IN STORAGE BOX ASSEMBLY, A3020113-001		1		
34		5995-01-182-2061	CABLE ASSEMBLY, RF (W9) A3020801-005	(80063)	STORED IN STORAGE BOX ASSEMBLY, A3020113-001		1		
35		5995-01-181-7082	CABLE ASSEMBLY, SP (W203) A3020817-001	(80063)	STORED IN STORAGE BOX ASSEMBLY, A3020113-001		1		
36			CABLE ASSEMBLY, SP (W204) A3020817-002	(80063)	STORED IN STORAGE BOX ASSEMBLY, A3020113-001		1		
37		5995-01-182-2054	CABLE ASSEMBLY, SP (W205) A3020818-001	(80063)	STORED IN STORAGE BOX ASSEMBLY, A3020113-001		1		
38		5995-01-182-2046	CABLE ASSEMBLY, SP-ELEC (W4) A3020800-001	(80063)	STORED IN STORAGE BOX ASSEMBLY, A3020113-001		1		

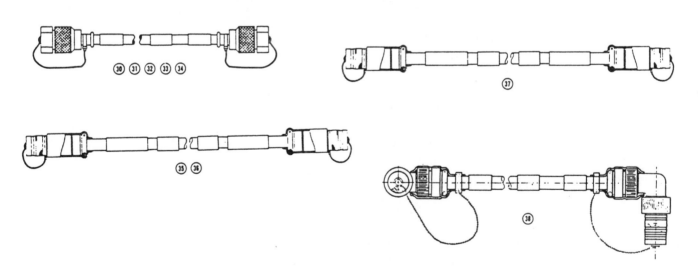

Figure 15-3 *An Example of a COEIL*

APPENDIX D

ADDITIONAL AUTHORIZATION LIST

Section I.   INTRODUCTION

D-1.   SCOPE.

This appendix lists additional items you are authorized for the support of the Antenna Group OE-361-(V)2/G (Single Carrier Feed) and the OE-361(V)2/G (Multiple Carrier Feed).

D-2.   GENERAL.

This list identifies items that do not have to accompany the Antenna Group OE-361(V)2/G or the Antenna Group OE-361(V)2/G and that do not have to be turned in with it. These items are all authorized to you byu CTA, MTOE, TDA, or JTA.

D-3.   EXPLANATION OF LISTING.

National stock numbers, descriptions, and quantities are provided to help you identify and request the additional items you require to support this equipment. The items are listed in alphabetical sequence by item name under the type document (i.e., CTA, MTOE, TDA, or JTA) which authorizes the item(s) to you. (Enter portions of next three sentences, only if applicable.) If the item you require differs between serial numbers of the same model, effective serial numbers are shown in the last line of the description. If item required differs for different models of this equipment, the model is shown under the "Usable on" heading in the desciption column.

Figure 15-4   *An Example of an Appendix Using an AAL*

APPENDIX D

Section II.   ADDITIONAL AUTHORIZATION LIST

(1) ITEM NO.	(2) LEVEL	(3) NATIONAL STOCK NUMBER	(4) DESCRIPTION	(5) UNIT OF MEAS
			PART NO AND FSCM	
			There are no additional authorized items required for the Antenna Group OE-361(V)1/G (Single Carrier Feed) or the Antenna Group OE-361(V)2/G (Multiple Carrier Feed).	

Figure 15-5   *An Example of an AAL*

APPENDIX E

EXPENDABLE SUPPLIES AND MATERIALS LIST

Section I.   INTRODUCTION

E-1.   SCOPE.

This appendix lists expendable supplies and materials you will need to operate and maintain the Antenna Group OE-361(V)1/ G (Single Carrier Feed) and the Antenna Group OE-361(V)2/G (Multiple Carrier Feed). These items are authorized to you by CTA50-970, Expendable Items (Except Medical, Class V, Repair Parts, and Heraldic Items).

E-2.   EXPLANATION OF COLUMNS.

a. Column 1—Item number. This number is assigned to the entry in the listing and is referenced in the narrative instructions to identify the material (e.g., ''Use cleaning compound, item 5, Appendiz D'').

b. Column 2—Level. This column identifies the lowest level of maintenance that requires the listed item. (Enter as applicable.)

       C - Operator/Crew
       O - Organizational Maintenance
       F - Direct Support Maintenance
       H - General Support Maintenance
       D - Depot

c. Column 3—National Stock Number. This is the National stock number assigned to the item; use it to request or requisition the item.

d. Column 4—Description. Indicates the Federal item name and, if required, a description to identify the item. The last line for each item indicates the Federal Supply Code for Manufacturer (FSCM) in parentheses followed by the part number.

Figure 15-6   *An Introduction to an Appendix Using an ES & ML*

(1) ITEM NUMBER	(2) LEVEL	(3) NATIONAL STOCK NUMBER	(4) DESCRIPTION	(5) U/M

Figure 15-7   *An Example of an ES & ML*

Chapter 16

The Glossary and Appendices in the Manual

Introduction

The glossary and appendices are supplementary to the text of the manual. The glossary can be placed after the introductory material to the manual, before the Introduction or Description Chapter, or at the end of the manual after the appendices. Appendices are always placed after the last chapter in the manual, and each appendix should be written as a separate unit.

Part I - The Glossary

A glossary is a special purpose dictionary containing definitions of unfamiliar terms used in the manual. If there are very few terms, they may be placed in the Introduction/Description Chapter. But if there are many terms to be defined, they must be placed in a glossary. You may find it necessary to repeat a few key terms from the glossary in the Introduction Chapter. In addition to unfamiliar terms, the glossary defines acronyms, non-standard abbreviations, and symbols used in the manual.

MIL-M-38784 stipulates, if a glossary is needed, "It shall immediately precede the index, if any. . . . " That document also specifies that a glossary may be used only when the terms are not adequately defined in the text of the manual.

In the manuals, you should use abbreviations and symbols in compliance with standards such as MIL-STD-12. But today technology advances at a faster rate than updates of standards. New terms and abbreviations for them are in "common use" long before changes incorporate them into the standards. The glossary, then, helps users keep up with technology.

Preparing the Glossary

When writing a glossary, you do not try to put every possible meaning you know into a glossary. A glossary is not a dictionary. Entries are based on the needs of the users; in other words, you would include only what users need to know to understand the manual.

There are two common formats for the glossary: modified paragraph and two column. In the modified paragraph format, you highlight the term by underlining, using all capital letters or different font types. The following is an example:

WIDOW—a single line or a single word carried over from the foot of one column or page to the top of the next.

In the two-column format, one column contains the terms and the other the definitions. If the glossary is sectioned, you may need to change the format of the two columns in each section. You may even combine the two formats, using the two-column format for abbreviations, acronyms, and symbols, and the modified paragraph format for the terms. Figure 16-1 shows a page from a typical military maintenance manual.

You do not need to write complete sentences in a glossary entry; just give the term and a few phrases which clearly define the meaning applicable in the manual. Usually, you do not need to put column headers on the glossary page. You can include acronyms, abbreviations, and symbols in the column of terms and show the words or the meaning in the definition column. If the glossary contains abbreviations, acronyms, symbols, and terms, you may need to divide it into sections. Figure 16-1 illustrates one sectioning concept.

Section I   Abbreviations (includes acronyms and symbols)

Section II   Terms

The format for glossaries is simplified, and the word "glossary" is centered at the top of the page, in all capital letters or highlighted by a different font type. Section titles are treated as major or center headers as shown in Figure 16-1, taken from MIL-M-63038B (TM).

Part II - The Appendices

An appendix to a technical manual performs the same function that an appendix performs in any good writing format. Each appendix contains information some users need but others do not. If you put this information in the body of the text, you would lose the interest of some users, so you place accessory information or "uncommon" text into an appendix for users who need that information. In operator manuals, the appendix material may include the following:

- References
- Various Types of Parts Lists
- Additional Authorizations Lists
- Expendables, Materials, and Supplies Lists
- Other Information Not Needed by Most Users

Typical Appendices

The contents of the appendix in the following military specifications are typical of the contents of current technical manuals:

MIL-M-63038B (TM)

TM 11-1520-232-20

GLOSSARY

Section I. ABBREVIATIONS

ac	Alternating current
adf	Automatic direction finder
af	Audio frequency
am	Amplitude modulation
ant	Antenna
A.P.U.	Auxiliary power unit
att	Attitude
aux	Auxiliary
bat	Battery (electrical)
bfo	Beat-frequency oscillator
B.I.T.E.	Built-in test equipment
dc	Direct current
fm	Frequency modulation
freq	Frequency
hf	High frequency
Hz	Hertz
ics	Interphone communication system
if	Intermediate frequency
iff	Identification friend or foe
inph	Interphone
max	Maximum
nav	Navigation
nbs	National bureau of standards
noness	Nonessential
omni	Omnirange
osc	Oscilloscope
ptt	Press to talk
rcvr	Receiver
rfi	Radio frequency interference
squel	Squelch
trans	Transmission
uhf	Ultra high frequency
Vac	Volts alternating current
var	Visual aural radio range
Vdc	Volts direct current
vhf	Very high frequency
vol	Volume
VOR	Vhf omnirange
xcvr	Transceiver
xmtr	Transmitter

Section II. DEFINITION OF UNUSUAL TERMS

Aft — At, near, or toward the rear of a helicopter; also the rear section of a helicopter.

Airframe — Structural components of a helicopter including the framework and the skin.

Figure 16-1 *Typical Glossary Page from Military Maintenance Manual*

- MIL-M-63036 (TM) Military Specification Manuals, Technical, Operator's, Preparation of
- MIL-M-63038 (TM) Military Specification Manuals, Technical, Organizational or Aviation Unit Direct Support, or Aviation Intermediate and General Support Maintenance

Preparing the Appendix

The information contained in the appendix of manuals specified by the two standards (MIL-M-63036 and MIL-M-63038) is consistent. Table 16-1 lists the contents of each appendix in these two specifications. Again, the customer's content specification will guide you in writing the appendices.

Appendix Format

If the customer has not specified a format, you could use your company's accepted format. For example, Figure 16-2 shows a typical appendix format including center headers, side headers, and page number.

The appendix designation and appendix title are centered headers. Letters and symbols in the center headers should

Table 16-1   *Comparing Appendix Content of Two Military*
Standards

Appendix	MIL-M-63036 (TM)	MIL-M-63038 (TM)
A	References	References
B	Components of End Item List Basic Issue Items List	Repair Parts and Special Tools List
C	Additional Authorization List	Maintenance Allocation Chart
D	Expendable Supplies and Materials List	Expendable Supplies and Materials List
E	none	Torque Limits
F	none	Marking Information
	Glossary	Glossary
	Index	Index

all be printed in upper case, centered on the print space of the page between the margins. You assign identifying letters (A) to each appendix and number the paragraphs with that letter as the first character. Number the pages sequentially for each paragraph using the identifying letter as the first character.

Conclusion

The glossary consolidates and defines special terms, abbreviations, and acronyms which are used in the manual. The glossary allows you to provide the reader with a specialized dictionary applicable to this particular manual. The appended material (appendices) allow you to provide more detailed information for some of the target readers which other readers neither need nor want to see. These two elements of the manual allow you to write a clear simple manual for all readers while incorporating more detail for a few readers. Some of the U.S. Armed Service specifications require specialized information in the appended material and put authorization for selected tasks into an appendix (such as the MAC, Maintenance Allocation Chart, in U.S. Army TM).

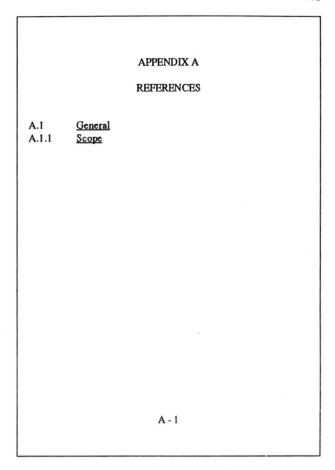

Figure 16-2   *An Example of an Appendix Page Format*

Chapter 17

Illustrations and Engineering Drawings in the Manual

Introduction

Illustrations are used in technical manuals to visualize complex information and hardware, equipment, etc. There are mainly two purposes for illustrations in manuals:

- To tell the operators and maintenance personnel what they need to know for effective, safe, proper use of the system.
- To help teach new operators and new maintenance personnel how to use the system.

A technical manual should have as many illustrations as necessary for the user to be able to understand the information and to perform the procedures correctly. Too many or too few illustrations can confuse the user.

Engineering Drawings and Military Standards

Engineering drawings come in all types and sizes. The drawings convey all the information a manufacturer requires to build an item. To be effective, engineering drawings must be properly organized, consistent, complete, and pertinent in detail. The following is a list of some of the Industry and Military Standards for drawings most frequently cited in government contracts.

ANSI-Y-10.19	Unit Symbols
ANSI-Y-14.14	Mechanical Drafting Practices
ANSI-Y-14.15	Electrical and Electronic Drafting Practices
ANSI-Y-14.17	Fluid Power Drafting Practices
ANSI-Y-32.2	Electrical and Electronic Graphic Symbols
ANSI-Y-32.3	Fluid Power Graphic Symbols
ANSI-Y-32.14	Digital (Logic) Graphic Symbols
ANSI-Y-32.6	Electrical and Electronic Reference Descriptions
DOD-STD-100	Engineering, Drawing Practices
DOD-D-1000	Drawings, Engineering and Associated Lists
MIL-STD-12	Abbreviations
MIL-STD-17	Mechanical Graphic Symbols

You should be familiar with these standards and when you want to use reproductions of engineering drawings in the manual, you need to check the specifications requirements. MIL-M-38784 stipulates guidelines as to what kind of engineering drawings are suitable in a manual, and DOD-D-1000 gives three drawing levels (listed below) that relate to equipment models and life cycle phases.

LEVEL	MODEL
1	Exploratory Development, Advanced Development, and Engineering Development
2	Engineering Development, Pre-production, and Production
3	Pre-production or Production

Level 1 drawings:	At a minimum discloses engineering design information sufficient to evaluate an engineering concept and may provide information sufficient to fabricate developmental hardware.
Level 2 drawings:	Discloses a design approach suitable to support the manufacture of a production prototype and limited production models.
Level 3 drawings:	Provides engineering definition sufficiently complete to enable a competent manufacturer to produce and maintain quality control of item(s) to the degree that physical and performance characteristics interchangeable with those of the original design are obtained without resorting to additional product design effort, additional design data, or recourse to the original design activity.

If you are preparing a technical manual that has stringent requirements for use of engineering drawings, you would not use Level 1 drawings and you might not even be able to use Level 2. Level 1 drawings may be quite rough and Level 2 drawings may not be properly prepared. For instance, MIL-M-38784, paragraph 3.6.26.1 states, "Parallel lines on wiring and schematic diagrams shall in no case be less than $\frac{1}{16}$ inch apart when reduced to print size." Your engineering drawing may be neat but the lines may not be properly spaced. Or, the printed characters such as reference designations or pin numbers may be so close to a line that when the drawing is reduced for the manual, "bleeding" (filling in) occurs. Or the symbols used by the design engineers may not agree with the symbol specification cited in the contract. Or, the names of assemblies on the engineering drawings might not match the names you are using as "official Nomenclature" or "common name." If the latter is true, you will have to modify the engineering drawing.

Types of Drawings

There are many types of engineering drawings (as defined in DOD-STD-100), but in the manual you will probably only use those listed in the Drawing List at the end of this chapter. Those with asterisks indicate the drawings most frequently used (see Figures 17-8 through 17-15). The Assembly Diagram, the Functional Diagram and the photograph are discussed in more detail in the following paragraphs.

● Assembly Diagrams

Assembly Diagrams are used to identify the location of parts within subassemblies, subassemblies within assemblies, assemblies within units or chassis, units within racks of equipment, and racks within entire systems. They can be direct reproductions of engineering drawings or used as source data to make into isometric or perspective drawings. They are used to show physical location (Figure 17-1 taken from MIL-HDBK-63038-1A (TM)), controls, and indicators, troubleshooting and removal and replacement. As an example, the Top Assembly drawing of a chassis shows all of the major assemblies and attaching parts. Its associated

MIL-HDBK-63038-1A (TM)

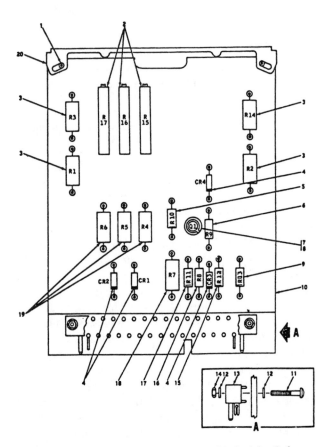

Figure 17-1 *Sample of Unit Numbering Method for Reference Designation*

Parts List is either made a part of the drawing (Integral Parts List) or is provided as a separate diagram.

The content of Parts Lists can vary. Typically, a Parts List contains the manufacturer's part number and Federal Supply Code for Manufacturers (FSCM) or Commercial and Government Entity Code (CAGE), quantity of each item, name and reference designation of each assembly, subassembly, or part, when applicable. The find numbers correspond to the find numbers in the field or the engineering drawing. If you were looking at a Parts List for item 10 on the Assembly XYZ, part number 1234567, reference designation A4, you would look for number 10 (usually located in a "bubble" or circle) on the engineering drawing. You would then find assembly A4 located within, or attached to the unit, which also would have a reference designation.

A reference designation is a series of numbers/letters used to identify everything from individual piece parts to systems. Reference designations of all assemblies take on their own and all other higher reference designations. For example, if assembly A4 in unit A2 were installed in rack 5 of several racks (1 through 5), the assembly's reference designation would be 5A2A4 (rack 5, unit A2, assembly A4). ANSI Y32.16 or IEEE STD 200 (Reference Designations for Electrical and Electronics Parts and Equipments) provides the requirements for reference designations.

● Functional Diagrams

Functional diagrams help the manual user understand how a piece of equipment works. First, they support the equipment theory of operation, both in its entirety and broken up into one or more smaller, functional diagrams. Second, they support the maintenance chapters by helping the user when troubleshooting a fault.

Functional diagrams show in either simplified or detailed block schematic, or logic diagrams how equipment works. Flow diagrams are generally laid out so equipment inputs (signals, power, etc.) are on the left side of the drawing and outputs are on the right side (Figure 17-2 taken from MIL-HDBK-63038-1A). This left to right layout gives the user a clear picture of how the unit works. Functional flow diagrams may or may not follow the boundaries of the hardware. At times you may have to show flow lines pointing toward the left. Complex circuits, such as feedback and closed loop circuits, often show flow lines in both directions.

Often you will have to show complex circuits in simplified form. A subassembly, such as a high density circuit card containing a hundred or more integrated circuits, may have a 10- or 15-sheet logic diagram. When writing the Theory of Operation Chapter, you may refer to that diagram, but you would normally make one, two or possibly three diagrams, each depicting a different level of detail, from that original 10 to 15-sheet diagram. The more detailed diagram would depict each major function, using symbols (schematic, logic, etc.).

You could make either simplified or detailed functional diagrams. Figure 17-3 taken from MIL-HDBK-63038-1A

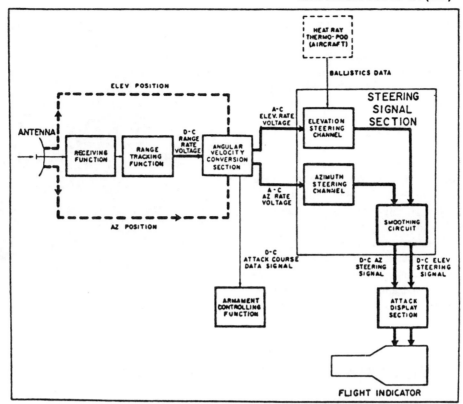

Figure 17-2 *Typical Functional Flow Block Diagram*

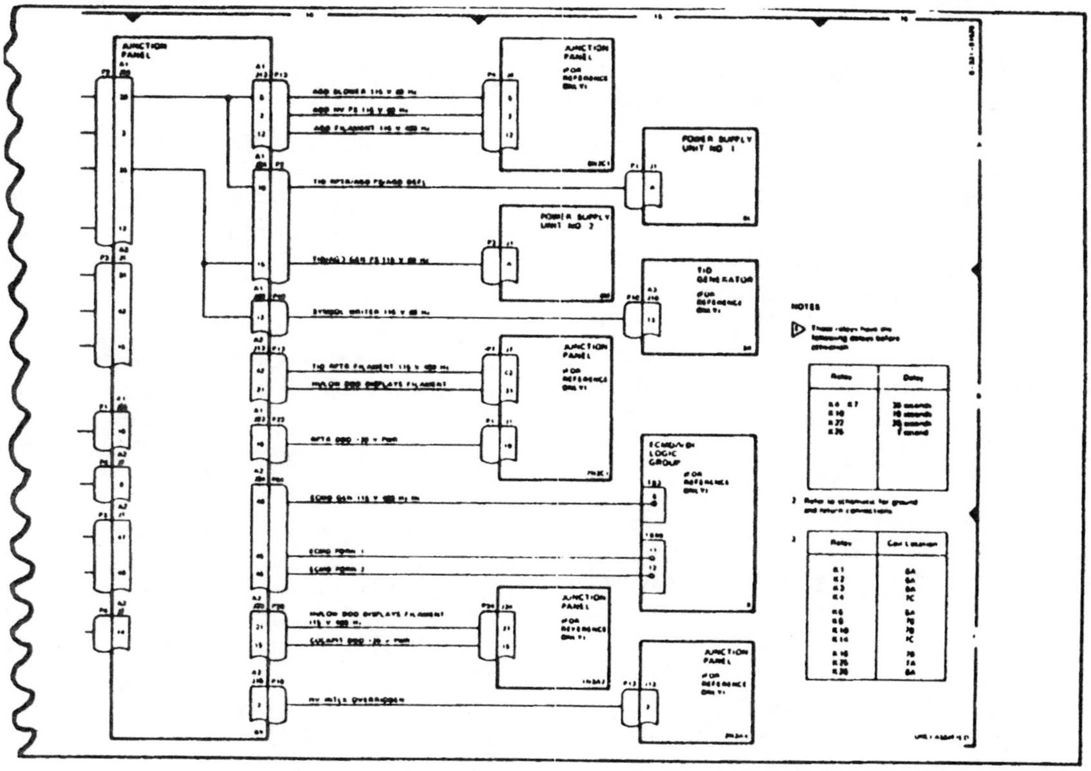

Figure 17-3 *Typical Power Distribution Circuit Diagram*

(TM), a Power Distribution circuit diagram, shows two levels of detail. Both levels show more detail than the block diagram shown in Figure 17-2. Figure 17-3 is a Power Distribution Circuit Block Diagram and shows the major functional assemblies. A Detailed Schematic Diagram shows the actual point-to-point wiring (called "pin-outs"). A wiring diagram of the power distribution circuit would provide all pin-out information.

● Photographs

It may be necessary to use photographs rather than line drawings to show various chassis, assemblies, etc. (See example of photograph, Figure 17-4). Today, the trend in government is to use line art. But at times you may need to prepare a manual not to Military Specifications or you may need to revise an existing manual which already contains photographs. If photographs are required, you need a photographer to take high quality, high resolution photos of the hardware. You should know from what angle the photos of the hardware are needed. For example, if you plan to use photos for Control and Indicator illustrations, you need front view photos. If you plan to use assembly and subassembly illustrations, you need shots of the hardware with the cover(s) removed. You will need to order at least two prints of each, one to be your "mark-up" copy and the other for the illustrator to make the final, boardmounted illustration. "Mark-up" means you indicate what callouts are required and what areas of the photo need to be deleted or highlighted.

Sizing Illustrations

When you prepare an illustration for a manual, you do not just start drawing. You must plan the size, type, layout and color of each illustration.

Drawings should be sized to fit the size of the manual. Most engineering drawings vary from A to J size. A standard 8-½ × 11-inch page is called an "A" size. Page sizes larger than the "A" size are multiples of the "A" dimensions as shown in Figure 17-5. "B" size drawings are twice as large as "A" size; "C" size is twice as large as "B," etc. Technical manual art is restricted to the vertical height of the page. Drawings larger than "A" size are folded at the unbound edges (top and outside edge). Drawings having a vertical height larger than the manual size ("C" size or larger) are called "foldout-foldup" art. Foldouts are printed on one side of a page, are always a right-hand page and are usually bound at the end of the book. Foldout art larger than B size is cumbersome and easily tears loose from the binders.

Paragraph 3.25 of MIL-M-38784 discusses the size limitations of foldout pages. The largest foldout permitted is 45 inches, including a full blank apron. The "apron" is an area of blank space at the bound edge and is as wide as the manual. The actual image area of the drawing starts at the

Figure 17-4 *Sample Photograph*

end of the apron. This blank area allows the user to look at the unfolded drawing while reading the text on another page.

The illustration itself is the actual usable image area of a drawing and is always less than the page size of the manual. Usable area means the space within the top, bottom and side margins. For example, an 8-½ × 11-inch manual page has a usable area of 7 × 9 inches.

Laying Out Illustrations

Some specifications give guidelines for laying out art on the page. Waveform and timing diagrams are examples of drawings which can be less than the size of the page. In the Army's "new look" manuals, page layout is very important. Color can also be used to highlight key text and art in the manual.

Any art for a manual must be sized, that is reduced, used as is, or expanded to dimensions which will fit on a full page size or partial page size. "Crop marks" at each corner of the illustration's perimeter identify the height and width dimensions. Paragraph 3.6.21 of MIL-M-38784 stipulates the following:

Each separately supplied illustration shall have the reproduction area defined by crop marks appearing on each side of the four corners marking the horizontal and vertical dimensions of the area to be reproduced. The lines shall extend no closer than ¼ inch to the outside of the reproduction area. The exact reproduction size shall be indicated between crop marks. Marks shall not be drawn with ball point pen or grease crayon. Crop marks shall be approximately ⅜ inch long and shall not cross or touch. (See Figure 3.1.)

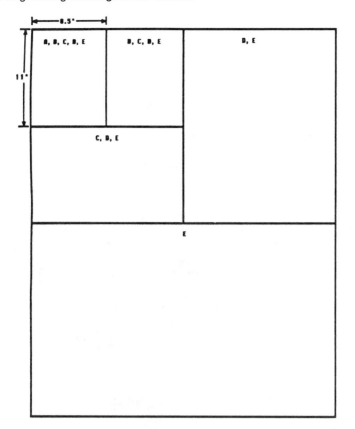

Figure 17-5 *Engineering Drawing with Size Relationships*

Figure 17-6, taken from MIL-M-38784B, illustrates crop marks on line drawings and photographs. The technique for generating illustrations is moving from being created as oversized board art (art generated by an illustrator on a drafting table) to digitized illustrations. Photographs can be digitized and changed, pixel by pixel, on the computer. Half tones or other screens are positioned by light pens, mouse, or graphics tablets. The single master photograph can be "exploded" or reduced to line art by simple machine processes. Line illustrations, such as block diagrams and test flow diagrams, are created on the computer more easily than by board illustrators. Computer illustrations can be done more and more by the technical writers nearly as well as by the illustrators. High resolution screens and magnification allow instant retouching, precise callouts, and effective linking of text, photographs, and graphics.

Boardmounted Art

Art is mounted on boards to preserve it. Paragraph 3.6.22 of MIL-M-38784 stipulates the following:

Continuous tone artwork that has not been prepared on illustration board shall be securely fastened to a mounting board by a process that will protect the artwork, will not discolor or distort the illustration, and will assure it being free of wrinkles and blisters.

Paragraph 3.6.24 of the same specification stipulates:

All boardmounted artwork shall be protected by an inner flap of non-oil tissue or vellum, and an outer flap of heavy paper. The tissue overlay and protective flap shall be cemented or securely taped to the back of the mounting board at the top of illustration and folded over the face of the illustration.

Figure 17-7 taken from MIL-M-38784B depicts the board mounting and covering of artwork (meaning art that is "mounted and flapped"). The illustration has crop marks and is sized and identified by a number (Art Control Number). An illustration number is added for identification and a flap is added to protect the surface of the illustration.

Conclusion

Technical manuals rely on various types of illustrations and engineering drawings to assist the users (equipment operators and maintenance personnel) in understanding their tasks. For the most part, manuals use assembly and functional diagrams, but photographs are used to a limited extent. Military Specification MIL-M-38784 gives guidelines on preparing illustrations. It also provides acceptable criteria for using engineering drawings and for preserving illustrations by boardmounting them.

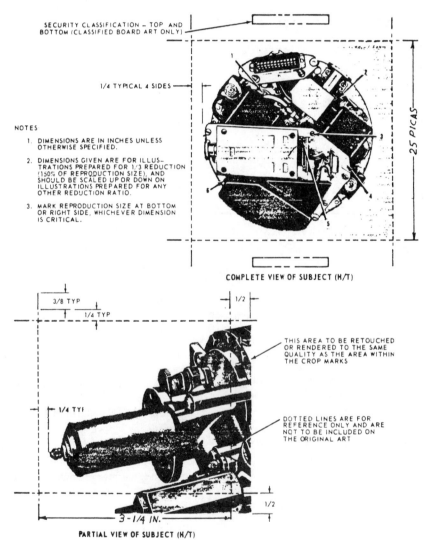

SECURITY CLASSIFICATION – TOP AND BOTTOM (CLASSIFIED BOARD ART ONLY)

1/4 TYPICAL 4 SIDES

25 PICAS

NOTES

1. DIMENSIONS ARE IN INCHES UNLESS OTHERWISE SPECIFIED.

2. DIMENSIONS GIVEN ARE FOR ILLUS-TRATIONS PREPARED FOR 1/3 REDUCTION (150% OF REPRODUCTION SIZE), AND SHOULD BE SCALED UP OR DOWN ON ILLUSTRATIONS PREPARED FOR ANY OTHER REDUCTION RATIO.

3. MARK REPRODUCTION SIZE AT BOTTOM OR RIGHT SIDE, WHICHEVER DIMENSION IS CRITICAL.

COMPLETE VIEW OF SUBJECT (H/T)

3/8 TYP   1/2
1/4 TYP

1/4 TYP

THIS AREA TO BE RETOUCHED OR RENDERED TO THE SAME QUALITY AS THE AREA WITHIN THE CROP MARKS

DOTTED LINES ARE FOR REFERENCE ONLY AND ARE NOT TO BE INCLUDED ON THE ORIGINAL ART

1/2

3-1/4 IN.

PARTIAL VIEW OF SUBJECT (H/T)

Figure 17-6 *Cropping and Sizing of Illustrations*

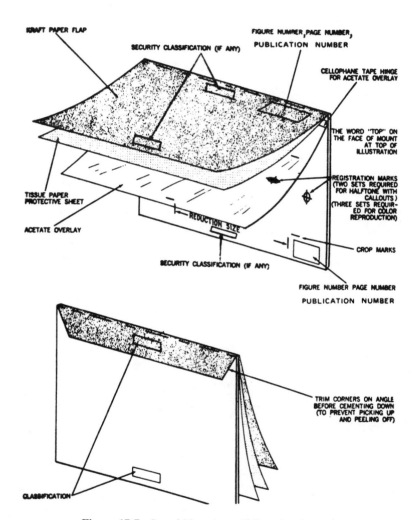

Figure 17-7 *Board Mounting and Covering Artwork*

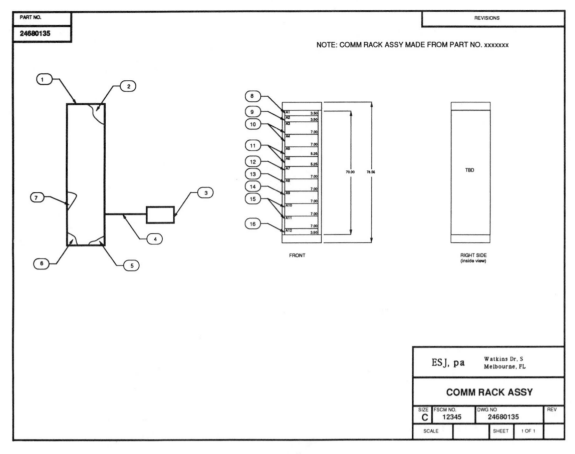

Figure 17-8 *Assembly Drawing*

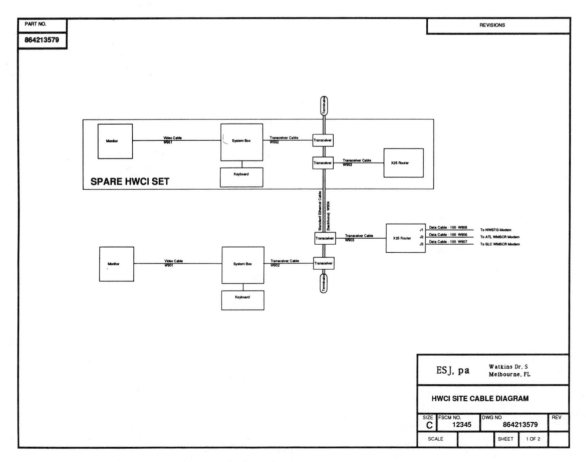

Figure 17-9 *Cable Schematic Example*

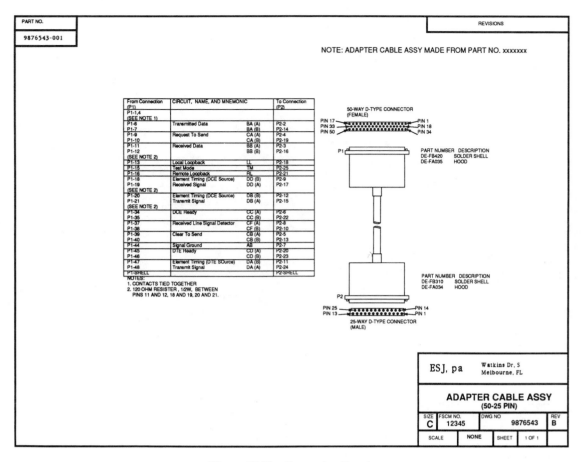

PART NO.

9876543-001

REVISIONS

NOTE: ADAPTER CABLE ASSY MADE FROM PART NO. xxxxxxx

From Connection (P1)	CIRCUIT, NAME, AND MNEMONIC		To Connection (P2)
P1-1,4 (SEE NOTE 1)			
P1-6	Transmitted Data	BA (A)	P2-2
P1-7		BA (B)	P2-14
P1-9	Request To Send	CA (A)	P2-4
P1-10		CA (B)	P2-19
P1-11	Received Data	BB (A)	P2-3
P1-12 (SEE NOTE 2)		BB (B)	P2-16
P1-13	Local Loopback	LL	P2-18
P1-15	Test Mode	TM	P2-25
P1-16	Remote Loopback	RL	P2-21
P1-18	Element Timing (DCE Source)	DD (B)	P2-9
P1-19 (SEE NOTE 2)	Received Signal	DD (A)	P2-17
P1-20	Element Timing (DCE Source)	DB (B)	P2-12
P1-21 (SEE NOTE 2)	Transmit Signal	DB (A)	P2-15
P1-34	DCE Ready	CC (A)	P2-6
P1-35		CC (B)	P2-22
P1-37	Received Line Signal Detector	CF (A)	P2-8
P1-38		CF (B)	P2-10
P1-39	Clear To Send	CB (A)	P2-5
P1-40		CB (B)	P2-13
P1-44	Signal Ground	AB	P2-7
P1-45	DTE Ready	CD (A)	P2-20
P1-46		CD (B)	P2-23
P1-47	Element Timing (DTE SOurce)	DA (B)	P2-11
P1-48	Transmit Signal	DA (A)	P2-24
P1-SHELL			P2-SHELL

NOTES:
1. CONTACTS TIED TOGETHER
2. 120 OHM RESISTER , 1/2W, BETWEEN PINS 11 AND 12, 18 AND 19, 20 AND 21.

50-WAY D-TYPE CONNECTOR (FEMALE)

PIN 17
PIN 33
PIN 50
PIN 1
PIN 18
PIN 34

P1

PART NUMBER	DESCRIPTION
DE-FB420	SOLDER SHELL
DE-FA035	HOOD

PART NUMBER	DESCRIPTION
DE-FB310	SOLDER SHELL
DE-FA034	HOOD

P2

PIN 25
PIN 13
PIN 14
PIN 1

25-WAY D-TYPE CONNECTOR (MALE)

ESJ, pa   Watkins Dr, S
Melbourne, FL

ADAPTER CABLE ASSY
(50-25 PIN)

SIZE	FSCM NO.	DWG NO	REV
C	12345	9876543	B

| SCALE | NONE | SHEET | 1 OF 1 | |

Figure 17-10   *Connection Drawing*

MIL-HDBK-63038-1A (TM)

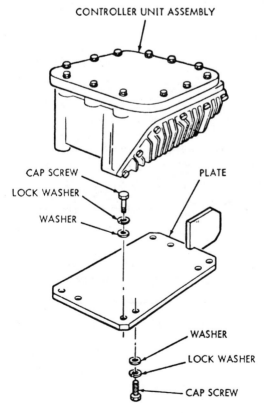

CONTROLLER UNIT ASSEMBLY

CAP SCREW

LOCK WASHER

WASHER

PLATE

WASHER

LOCK WASHER

CAP SCREW

Figure 17-11   *Example of an Exploded Drawing*

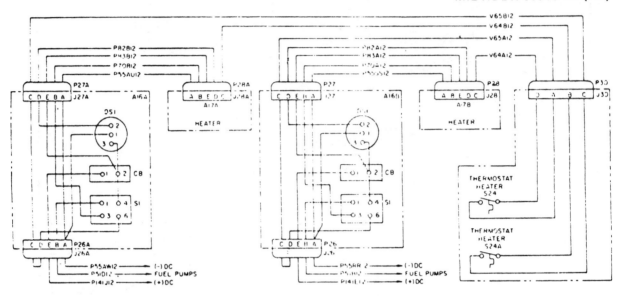

Figure 17-12 *Interconnection Drawing*

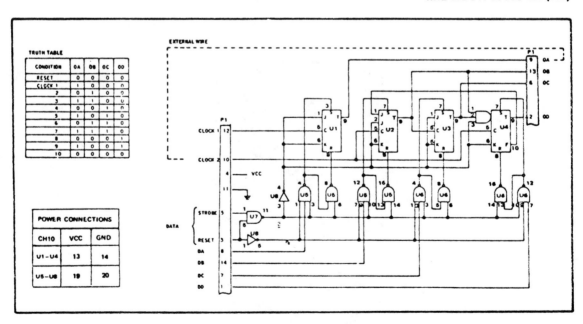

Figure 17-13 *Logic Drawing*

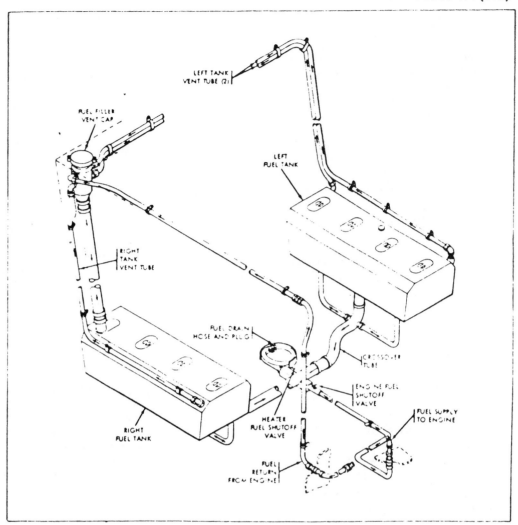

Figure 17-14 *Example of a Piping Drawing*

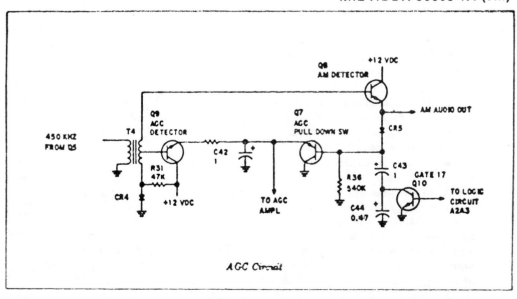

Figure 17-15 *Schematic Drawing*

Drawing List

Drawing	Purpose
Altered Item	Delineates complete details of the alteration. When any completed item is to be altered, the designers responsible for the alteration shall prepare an altered item drawing. The drawing includes necessary information to identify the item prior to its alteration including the original identifying part number and, if a commercial or vendor-developed item, the name (address if known), and manufacturer FSCM number of the source of the original part.
Assembly*	Depicts the assembled relationship of (a) two or more parts, (b) a combination of parts and subordinate assemblies, or (c) a group of assemblies required to form an assembly of higher order. (See Figure 17-8.)
Cable Assembly*	Depicts power, signal, radar frequency, or audio frequency cables normally used between equipment, units, inter-racks, etc. Cable terminations are normally plugs, sockets, connectors, etc. (See Figure 17-9.)
Connection or Wiring*	Shows the electrical connections of an installation or its component devices or parts. It may cover internal or external connections, or both, and contains such detail as is needed to make or trace connections that are involved. A connection diagram usually shows general physical arrangement of the component devices or parts. (See Figure 17-10.)
Control*	Discloses configuration and configuration limitations; performance and test requirements; weight and space limitations; access clearance, pipe and cable attachments, etc., to the extent necessary that an item can be developed or procured on the commercial market to meet the stated requirements; or, for the installation and co-functioning of an item to be installed with related items. Control drawings are identified as Envelope, Specification Control, Source Control, Altered Item, Selected Item, Interface Control, or Installation Control drawings.
Detail	Depicts complete end-item requirements for parts delineated in the drawing except when additional end-produce requirements are accomplished on inseparable assembly drawings, e.g. mating holes.
Diagrammatic	Delineates features and relationship of items forming an assembly or system by means of symbols and lines. A diagrammatic drawing is a graphic explanation of the manner by which an installation, assembly or system (e.g., mechanical, electrical, electronic, hydraulic, pneumatic) performs its intended function.
Elevation	Depicts vertical projections of buildings or structures or profiles of equipment such as aircraft, automotive, and marine or portions of the same.
Envelope	Depicts an item, either government or privately developed, where it is desirable to have all features, other than those shown on the drawing, left to the ingenuity of the producer to meet the specified performance data and design requirements.
Exploded Assembly*	Using either isometric or perspective drawing techniques, depicts the individual items that make up a part in a manner whereby they are separated from each other but related to each other by the use of a center line. (See Figure 17-11.)
Installation	Shows general configuration and complete information necessary to install an item relative to its supporting structure or to associated items.
Installation Assembly	Shows the installed and assembled position of an item(s) relative to its supporting structure or to associated items.
Installation Control	Sets forth information for an item in terms of area, weight and space, access clearance, draining clearances, pipe and cable attachments required for the installation and co-functioning of the item to be installed with related items. Control drawings include the following information as applicable: overall and principal dimensions in sufficient detail to establish the limits of space in all directions required for installation, operation and servicing; the amount of clearance required to permit the opening of doors or the removal of plug-in units; clearance for travel or rotation of any moving parts, including the centers of rotation, angles of train in azimuth, elevation and depression, and radii from each pivot to the end of each rotating element involved in clearance determination.
Interconnection*	Is a form of connection or wiring diagram which shows only external connections between units, sets, groups, and systems. (See Figure 17-12.)
Interface Control	Is a statement of the physical and functional interface between items which must function together. Includes engineering requirements, coordinating data, compatability needs, signal names, and function, size, fasteners, connectors, location, limitations, and controls. Is used to ensure that changes to one item do not create a functional problem for the total operation.
Kit Drawing	Indicates or depicts a packaged unit, item, or group of items, instructions, photographs, or drawings which are used in modification, installation, or survival. The items in the kit normally do not in themselves constitute a complete functional assembly. A Kit Drawing may be a listing of part numbers, a pictorial representation of parts, or a combination of both.
Logic*	Shows by means of graphic symbols the sequence and function of logic circuitry. (See Figure 17-13.)

Modification — Delineates changes to delivered item assemblies, installations, or systems. Modification drawings are prepared to add, remove, or rework item equipment installations, or systems to satisfy the users' requirements or to incorporate mandatory changes (i.e., safety, reliability, or application extension) in delivered equipment.

Piping* — Depicts (in a hydraulic, pneumatic, or fluid diagram) the interconnection of components by piping, tubing, or hose, and when desired, sequential flow of fluids in the system. (See Figure 17-14.)

Plan — Depicts a horizontal projection of a structure, showing the layout of the foundation, floor, deck, roof, or utility system.

Printed Wiring Master Drawing — Shows the dimensional limits or grid location applicable to any or all parts of a printed wiring or printed circuit including the base.

Printed Wiring Master Pattern — Is a precise scale pattern which is used to produce the printed circuit within the accuracy specified in the printed wiring master drawing.

Schematic* — Shows, by means of graphic symbols, the electrical connections and functions of a specific circuit arrangement. A schematic diagram facilitates tracing the circuit and its functions without regard to the actual physical size, shape, or location of the component devices or parts. (See Figure 17-15.)

Selected Item — Defines an existing standard or design or vendor activity item with further required selection or restriction of the item for fit, tolerance, performance, or reliability within the range or limits prescribed for that item.

Source Control — Depicts an existing commercial or vendor item which exclusively provides the performance, installation, and interchangeable characteristics requires for one or more specific critical application.

Specification Control — Depicts an existing commercial item or vendor-developed item advertised or catalogued as available on an unrestricted basis on order as an "off-the-shelf" item or an item, while not commercially available, that is procurable on order from a specialized segment of an industry.

Wiring Harness — Shows the path of a group of wires laced together in a specified configuration, so formed to simplify installation.

Wiring List — Is an engineering drawing consisting of tabular data and instructions required to establish wiring connections within or between unit or equipment, or between equipments, sets or assemblies of a system. A wiring list is a form of interconnection or connection diagram.

Part III

THE SUBCONTRACTING PROCESS AND COMPUTER-AIDED ACQUISITION AND LOGISTICS SUPPORT

Chapter 18

Subcontracting Technical Publications

Introduction

When it comes to technical publications, subcontracting is common practice. Why do prime contractors decide to have all or part of the technical publication tasks done by someone outside the company? Several reasons make subcontracting an attractive alternative to preparing them in-house:

- Availability of personnel
- Hardware complexity
- Availability of engineering data
- Customer needs
- Cost
- Schedule
- Availability of capable subcontractors

Choosing a Subcontractor

Choosing the right subcontractor depends on evaluating such factors as cost, personnel, and ability to perform the task. Depending on the management's, technical writer's or subcontractor's point of view, subcontracting the writing task can be either negative or positive. You, as a technical writer, need to understand all sides of the issue. There is, however, an element of risk to subcontracting. Subcontractors are in business to make a profit, so a well-written subcontract and a strong subcontract administrator are crucial to obtain a successful final product. The following is a list of some of the main reasons to subcontract when:

1. Government classification does not preclude use of a subcontractor.
2. In-house personnel capability has reached saturation and long term prospects do not indicate a continuing trend of increased work loads (a good downstream work load may be justifiable grounds for increasing staff size).
3. The schedule cannot be met by in-house capability.
4. A cost problem exists and combined subcontractor/in-house administration costs will be lower than the cost of accomplishing the task in-house.
5. Availability of a complete data package (equipment baseline, drawings, test procedures, supporting vendor data, in-house engineering documentation such as a Critical Design Review (CDR) package, Logistics Support Analysis (LSA), data etc.) will permit a subcontractor to work effectively and efficiently.

6. All source data (e.g., engineering drawings and Engineering Change Orders (ECO's)) required by the subcontractor can be completed before the "freeze date."
7. A reliable schedule is developed that won't slip (to prevent a "change in scope" from the subcontractor).
8. A pre-emptive decision to subcontract is made during the proposal stage as a means of enhancing the company's bid price.
9. Use of "body shop" labor is not considered a viable solution to task completion.
10. A customer has not indicated a negative attitude towards subcontracting.

Management's View

Management represents the interests of the company and the stockholders. Management is primarily interested in three aspects of any Technical Publications contract: cost, schedule, and quality of the final product. Likewise, the technical publications manager and co-workers (technical writers, parts listers, editors, illustrators, and production personnel) are concerned with cost, schedule, and final product.

Management's goals are to satisfy the customer and make a profit. It attains these goals by building a solid team of workers. If a company subcontracts all of its writing to subcontractors, it will need only a few workers to oversee the subcontractors' work and ensure it meets the contract requirements. As a result, the company will have a team of subcontract administrators and an almost nonexistent in-house technical writing staff. Some companies prefer to do without a publications staff, while other companies prefer to have a staff, even a minimal staff, to handle publications. Management will normally staff up to that level where normal work loads are easily handled. Peak work loads can be handled by having employees work overtime, by hiring job-shoppers or by subcontracting. This gives the company the ability to spread out staff at low periods of work rather than laying them off.

Technical Writer's View

The technical writer relies on management to provide enough work to justify the writer's job. You, as a writer, must learn the company's way of doing business and be able to interact well with company customers and company employees, both vertically (management) and horizontally (your co-workers). For every task you face there is a learning curve. Every new contract brings new requirements (specifications), so you need to keep pace with new technology to grow professionally and to be competitive.

If the technical publications task is subcontracted, the technical writer assigned to the task becomes a subcontract administrator. This subcontract administration effort should not be confused with the job of a Subcontract Administrator who works for the company's Subcontract Administration function. The writer as subcontract administrator is given no legal power, but is rather a monitor of the subcontractor's efforts. This subcontract administration cost must be added to the cost of the subcontract to obtain the final cost of the task. As a result, the combined expenses of a subcontract task plus subcontract administration costs can be almost equivalent to the cost of doing the same task within the company.

Subcontractor's View

Subcontractors exist because they can perform a contractor's task quickly, cheaply, and reliably. For reasons stated previously, a contractor may wish to subcontract a writing task. Subcontractors are almost totally dependent upon "good times" when government spending is high (i.e. awarding defense contracts). There are basically four types of subcontractors: job houses, prime contractors, body (job) shops, and freelance writers.

● *Job Houses:*

A typical technical publications job house employs sales people, technical writers, illustrators, parts listers, editors, and clerks in addition to managerial and accounting personnel. The editorial staff usually consists of one or more editors and proofreaders. Quite often their business is only preparation of technical data, which limits their ability to diversify. The job house may be classified as a small business. It may own the capital equipment needed to produce technical manuals. Many job houses use subcontractors for services such as photography and printing. Contracts are obtained by the sales staff who, through research and business ties, develop a list of potential customers through various publications such as the *Commerce Business Daily,* which is "a daily list of U.S. Government procurement invitations, contract awards, subcontracting leads, sales of surplus property and foreign business opportunities."

People who work for job houses are usually high quality, experienced professionals. They are used to dealing with a wide variety of customers, and have extensive experience writing to many different specifications. With this flexibility they can get a task done with fewer people.

● *Prime Contractor as Subcontractor:*

Some prime contractors allow their technical publications staff to subcontract their services to other prime contractors. These writers usually cost more than those of job houses because of higher overhead rates. Consequently, the work done by the prime subcontractor may have a higher cost.

Also a job house may be more willing to accept lower profit than the prime contractor in order to keep the staff busy during low periods of work.

● *Body (Job) Shop:*

A body (job) shop is similar to an employment agency. The shop acts as a labor source for contractors needing short-term help for specific tasks. The technical writer in a body shop is given higher wages to compensate for lack of company benefits (no vacation, no sick pay, no hospitalization, etc.). It is not unusual for a writer's name to be on file with several body shops. Writers working though body shops move often, even taking assignments overseas. For this short-term work there would be a contract between the prime contractor and body shop and between the body shop and the writer. The total cost of body shopper's wages to the writer and their fee may be the same or higher than that paid by the contractor to his/her own writer, especially if the contractor must add on a higher overhead and General and Administrative (G&A) costs. Using the service of body shop labor helps contractors in that the need for the part-time writer ends when the job ends, and the person can be released.

● *Freelance Writers:*

Freelance writers sell their services to contractors and subcontractors while working out of their own offices or homes. Because of limited funds for capital equipment they may lack all the technical publishing equipment. Freelance writers may provide only a draft manuscript and rough art as their deliverable product and not camera-ready copy or final art. As a result, the contractor or subcontractor is responsible for generating final art and printing the manual. With the advent of home computers these writers now are capable of providing their "camera-ready copy" in electronic media form through desktop publishing.

The freelance writer is a back-up to the subcontractor who is pressed for time or overloaded with work. Like job shoppers, freelance writers work on direct contracts with prime contractors and may be given needed work space and equipment at the contractor's location. Their cost is usually lower than that of body shop labor because of their lower overhead.

The Subcontract Sequence

Figure 18-1 is a flow diagram of 40 blocks which depict the various stages in a subcontract process.

Contract Award

Block 1 represents the process of the contractor preparing a proposal in response to a customer's Request for Bid, Request for Quote, or Invitation for Bid. This process is the same for the subcontractor as it is for the contractor (Chapter 4).

Make/Buy Decision

Block 2 represents the contractor's make or buy decision. The decision is based on factors including cost, schedule, product quality, data, and available personnel. A "NO" decision indicates the task should be done in-house (Block 3). A "YES" decision (Block 4) starts the investigation process to review the size of the task. A decision to subcontract can be made during a contractor's proposal stage rather than after the contract has been awarded. An early buy decision is usually based on the cost or the schedule of the writing task.

Subcontractor Selection

At this point, the technical publications manager becomes involved with investigating potential subcontractors by considering the following:

1. Previous experience with one or more subcontractors.
2. Research through an on-hand file of subcontractor brochures (sent out by subcontractors to contractors who may be potential customers).
3. Word of mouth references from other technical publications people who either have used (or worked for) subcontractors, or have not used them but know of their existence and reputation.

Potential subcontractors are divided into two categories: "known" and "unknown." Those in the "known" category may not require a vendor survey. In other words, it may not be necessary to visit the subcontractor. However, if a subcontract is large enough, upper management may insist on sending key personnel (contract, legal and/or financial representatives) to investigate the subcontractor. A vendor survey includes a financial check and a visit to the facilities. Vendor surveys cost the contractor money, especially if the subcontractors are geographically diverse. It is therefore desirable to restrict the number of subcontractors who are to be visited. By waiting until the proposals are received and evaluated, one or more of the subcontractors may be initially disqualified on the basis of inadequate staff, technical incompetence, or high cost.

Sometimes a subcontractor investigation (Block 5) will take place after the Bidders Conference is held. Unknown subcontractors need to be thoroughly investigated. Their facilities, employees, capital equipment, financial status, and reputation should be closely reviewed to find out the following:

1. Is their facility physically large enough to handle the task?
2. Is their staff (technical writers, editors, proofreaders, quality controllers, illustrators, typists/word processors, parts listers) large enough to produce the data in the time required by the contract? If not, are they going to rely on subcontractors or freelance writers to help them carry the load?
3. Do they have a qualified staff (technically capable and have previous experience with your customer; with knowledge of the cited military specifications; with knowledge about the type of equipment you are manufacturing)?
4. Are they financially capable of sustaining contractually incurred expenses, or do they require closely spaced, periodic payments?
5. Have they previously (and successfully) handled tasks of this cost and size?
6. How close is their annual budget to the total contract price and are they overestimating their available assets?
7. Do they have a reputation for "change of scope" tactics?
8. Do they have their own capital equipment (cameras, enlargers, photo lab, printing press, etc.) or are they dependent upon sub-subcontractors for this work?
9. Are they physically located within an acceptable proximity to your company so they can easily travel back and forth for information gathering, or will they have to pump up the cost of the contract for travel expenses? If they are distant and the contract is a large investment, will they have a representative or a staff permanently located in your area for the duration of the contract?
10. If the contract is classified, is their facility secure and does the staff possess proper clearances?

Data/Package Statement of Work (SOW)

The two most crucial aspects of a technical publications subcontract are the data package and the SOW. Both are presented at the Bidders Conference (Block 6). When preparing their proposals, subcontractors place heavy stress on the accuracy and adequacy of the data package containing all pertinent equipment information. The SOW must define all tasks to be performed, every deliverable and milestone. Failure to do so could result in problems later on in the project.

Bidders Conference

At a Bidders Conference (Block 7), the technical publications manager usually acts as the conference moderator. The manager introduces everyone, distributes handouts (such as the revised agenda), identifies all deliverables, and runs through the SOW. What tasks the contractor and the subcontractor are each responsible for are stressed.

The engineering representatives give a technical description of the equipment using handouts, viewgraphs, slides, etc. They review all the equipment configuration, operation, and maintenance, so the subcontractor can better evaluate the data package.

Subcontractors are invited to survey the data package, either separately or jointly. The contractor usually will not make a separate data package for each bidder unless the package is small. Each subcontractor may be given a set time period to examine the package privately or the contractor may decide to lay the package on the conference table to let all subcontractors survey the data one at a time.

All subcontractors can ask the contractor representatives

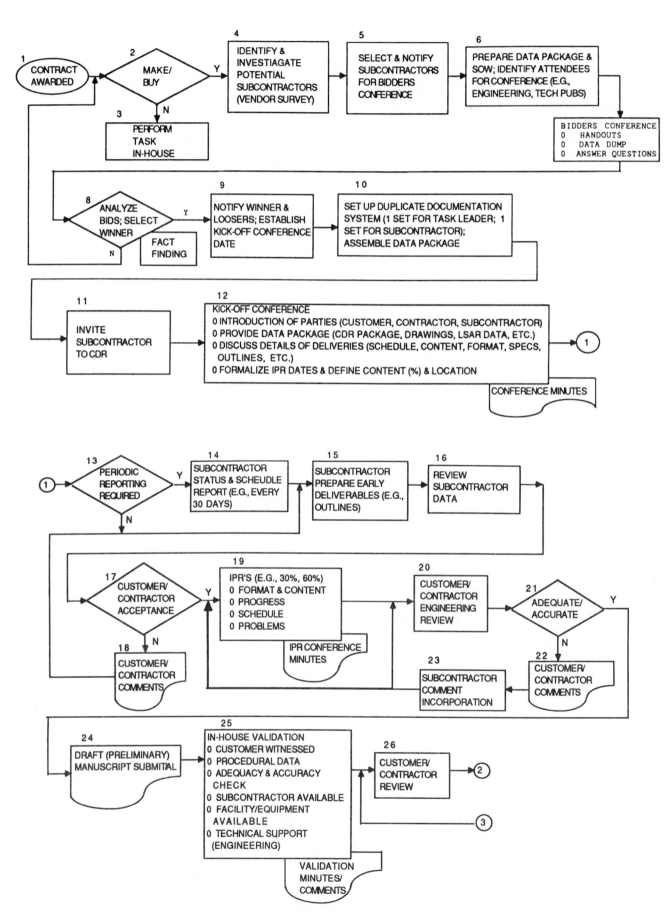

Figure 18-1   *Subcontract Administration Flow Diagram*

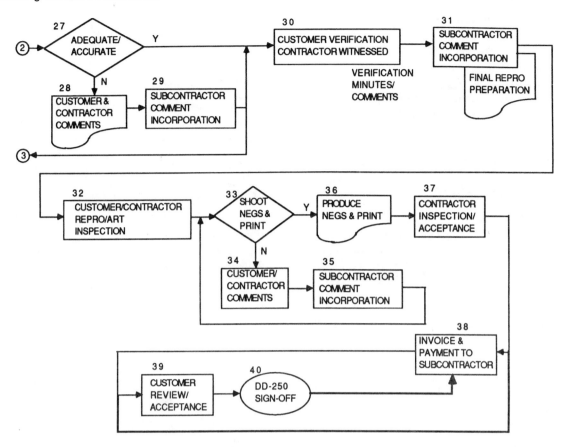

Figure 18-1 *Continued*

questions. The contractor must convey all information to the subcontractors equally. Failure to do so could give the impression of favoritism and/or bid rigging. The losing subcontractors could press charges against the contractor, and this can have serious consequences, especially when government contracts are involved. Two government agencies within the Department of Defense, the Defense Contract Audit Agency (DCAA) and the Defense Contract Administration Service (DCAS), are set up to administer and monitor government contracts. Government agents have the power to impound materials and to bring charges against contractors suspected of rigging prices, etc. This can affect both the company and their employees through fines, loss of contracts, loss of security clearances, and even imprisonment.

Subcontractor Proposal Evaluation

Block 8 represents the evaluation of subcontractor proposals. Subcontracts are evaluated by the technical publications manager, task leader, subcontract manager, possibly a finance representative, the Integrated Logistics Support (ILS) manager, and the program manager. These evaluations are reviewed to find the best proposal. This group may have the sole authority to award the contract or it may make a recommendation to the Program Management Office

(PMO). The subcontractors who are the finalists may be called for further negotiations before the winner is selected.

The contractor, after reviewing all the data (Block 8), determines which subcontractor has the best proposal. The winner is notified (Block 9) and a kick-off conference date is set. The task leader then sets up a duplicate documentation system that will be the same as the subcontractor's (Block 10). By using this system, the subcontractor and contractor possess the same data and receive updated data at approximately the same time, and can discuss any part of a document that is confusing. If the CDR has not yet been held, the subcontractor should be invited (Block 11) to get more information on the hardware operation and configuration.

A kick-off conference date is set for the customer, contractor, and subcontractor to exchange information (Block 12). In some cases, if the contractor does not want the subcontractor to interface directly with the customer, the contractor may not invite the subcontractor to this conference, but will set up another separate kick-off conference. The subcontractor is given the source data for deliverable requirements and the manual's format and content requirements. In-Process Review (IPR) dates and locations may also be set up, and the manual's review cycle milestones are identified (30 percent, 60 percent, etc.).

After the conference, all meeting minutes are signed by the key members attending. The contractor and subcontractor may be given up to a month to respond to the customer's comments. No response means automatically accepting all comments as stated in the minutes.

The subcontractor, depending upon the contract, may or may not be required to make periodic reports (Block 13). There are times when a customer will require (by way of a Contact Data Requirements List (CDRL)) the contractor to make a monthly status report. This requirement may be passed onto the subcontractor. If it is (Block 14), the subcontractor will prepare monthly progress reports.

Subcontractor Start-Up

Some deliverable items are required early in the program. In addition to the outlines, deliverables can include Maintenance Allocation Chart (MAC) and Ground Support Equipment Recommendation Data (GSERD), or Support Equipment Recommendation Data (SERD) (Block 15). The MAC and GSERD are usually by-products of the contractor's LSA and LSA Record data. If the subcontractor is given the task to prepare these deliverables, the task will may be limited to putting the documents into proper format. The customer reviews the data to determine if the recommendations are acceptable (Blocks 16 and 17), and the customer and contractor review the comments (Block 18). The subcontractor must incorporate the comments, and the documents are resubmitted to the customer for approval.

In-Process Reviews

Since the prime contractors are not actually performing the writing task, they and the customer rely on the subcontractor for the In-Process Reviews (IPR's). Prior to the first IPR (Block 19), only the subcontractor actually knows how the manual is progressing.

The subcontractor must make copies of the manual available to all the reviewers. Depending upon the size and importance of the task, the number of people present can vary. Customers (the buyers) may send only one person or they may send 10 or more reviewers. If the task is tri-service, each interested branch of the Department of Defense (Army, Navy, Air Force) may send representatives. The prime contractor may send just the Publications Manager, or may include the ILS Manager, and Lead Tech Writer assigned to manage the subcontractor's task.

A wrap-up conference is held at the end of the meeting. Minutes of the IPR are signed. The subcontractor by this time understands exactly what the customer wants and knows the best way to approach the rest of the technical publications task. By the next IPR, all comments from the previous IPR have been responded to, and more material should be available for this review. This process is repeated at every review until the preliminary manuscript is ready for validation and verification.

The subcontractor submits the manuscript to the contractor so (Blocks 20 and 21) the data's accuracy and adequacy can be reviewed. The prime contractor's hardware design engineers should receive a copy of the manual to check its technical accuracy. The subcontractor then must incorporate all comments into the documents (Blocks 22 and 23).

Preliminary Manuscript Review

A final conference may be called (sometimes referred to as a 100 percent IPR), and the completed manuscript is made available to the customer (Block 24). This is the final review before the validation process when the manual is compared against the hardware. At this point, neither the customer nor the contractor nor the subcontractor expect the manual to be 100 percent accurate. The validation process helps find the errors so the corrections can be made (Block 25).

Validation/Verification

The process of validation and verification (Blocks 26-32) is the same for the subcontractor as it is for the contractor as described in Chapter 8.

Final Deliverable Preparation

After incorporating all customer and contractor comments into the manuscript, the subcontractor makes a last inspection of the material, text, tables, and art work (Block 32). If the material is accepted, the customer permits the negatives (if deliverable) to be shot (Block 33). If the customer finds errors or poor quality in the repro, the subcontractor will have to make the necessary corrections (Blocks 34 and 35) before submitting an invoice for payment. After the last corrections are completed, negatives are made, and a number of copies of the manual are printed as required by the CDRL (Block 36), and a few "house copies" are usually made for the contractor's archives.

Repro Delivery/Final Acceptance

The subcontractor task is not necessarily over when the last corrections are made to the manual and the final repro (camera-ready copy) is sent to the customer (Block 37). The SOW may state that the customer, not the contractor, must give final approval of the technical manual. The customer verifies accepting the manual by signing-off on Form DD-250, which makes the deliverable item billable (Blocks 38-40). Depending on the requirements of the contract, the deliverable may be reproducible material, or original art, or negatives, or printed copies of the manual, or computer disks. Final payment is made to the subcontractor only after the customer has approved final delivery.

Contractor's Rights and Obligations

The contractor has rights only within the limitations of that specific task performed by the subcontractor. A subcontractor cannot be asked to do anything that lies beyond the realm of that task. Some contractor's rights include the following:

1. The right to alter the subcontract as necessary to get the task done. Any modification of the subcontract must be with subcontractor concurrence. An example of subcontract modification would be an increase or decrease in task size due to hardware configuration change, which would increase or decrease the amount of work the subcontractor has to perform. It is recognized that there are limits to what can be expected in terms of a subcontractor's ability and obligation to perform.

2. The right to periodically inspect progress and to accept or reject the quality or quantity of work. Limitations exist where the subcontractor's proposal states clearly defined limits, such as page counts or depth of coverage.

3. The right to cancel a contract if the subcontractor is not performing. This right has many ramifications and is used only when absolutely necessary. Cancellation of a contract can take a long time and may require extensive proof showing lack of performance.

4. The right to accept or reject the subcontractor's selected sub-subcontractors to assist the subcontractor, providing that the contractor has clearly defined this right in the subcontract.

5. The right to expect the subcontractor to give warning in advance should problems exist that will prevent completion of the subcontract in a timely manner.

Some contractor's obligations and limitations include the following:

1. Changes to requirements for the manual content must be restricted to a reasonable percentage of change. Changes brought about by hardware, software, or data changes may be picked up ''gratis'' by the subcontractor to show willingness to perform. However, the subcontractor cannot be expected to rework material over and over again with no cost impact to the contractor. The percent change (percent of the total dollar value of the subcontract) considered acceptable will vary depending upon the nature of the change and the overall size of the subcontract. It is reasonable to expect the subcontractor to pick up normal minor design changes up to 30 or 60 days prior to completion of the draft manuscript delivery (the ''freeze date'').

2. The contractor does not have the right to select the subcontractor's staff performing the writing task other than those key individuals identified in the subcontractor's proposal. Nor can the contractor dictate the number of writers to be used on the task if not specified in the contract.

3. The contractor is obligated to provide the subcontractor with information in a timely manner so that schedules can be met.

4. The contractor must permit the subcontractor to have access to design engineers and to hardware as necessary to obtain information in addition to that provided on the engineering drawings.

5. The contractor should give the subcontractor adequate warning when planning a facility visit or when requiring the subcontractor to visit the contractor's facility.

Conclusion

Companies can fulfill their technical publications contract requirements by using both in-house resources and subcontractors. There are numerous factors involved in selecting subcontractors and monitoring their output, but the sequence of events in subcontract performance parallel those of the prime contractor's performance milestones. The keys to establishing a viable subcontracted technical publications task are based on preparing a solid SOW and selecting a reliable subcontractor. Writers given responsibility for administering the subcontractor's work must understand the extent of their control and have the ability to make decisions. Since the subcontractor is considered an extension of the prime contractor, all personnel who interface with the prime contractor's customers and all work done by the subcontractor reflect directly on the contractor's performance. The competent subcontract administrator must pay close attention to all aspects of preparing data to assure quality technical publications.

Chapter 19

Data-Based Technical Manuals

Introduction

A revolution known as "desktop publishing" is underway in the world of technical manuals. Through computer technology one person can handle the writing and production of technical manuals when once it took the skills of numerous specialists. Technical writers now can write, edit, spell-check, check for readability, modify, illustrate, layout, and print an entire manual using a computer and various sophisticated software packages. Word processing programs make it possible to generate text and tabular material set to specific style guides, automate spelling checks, generate the table of contents, and insert illustrations from other documents. Illustrating programs can generate both two and three-dimensional drawings. Color can be used to highlight text, tables, and illustrations.

Today there is a need for writers to write technically accurate data to the format and content requirements of customer or government specifications using "desktop publishing." The need to change from manually-generated to computer-generated technical manuals is no longer an option. Competition and customer needs are mandating the use of this new technology.

Technical Data and the Department of Defense

The government has identified a plan for putting all data developed for weapons systems it receives from contractors into one comprehensive data base system. The system is referred to as Computer-aided Acquisition and Logistic Support, or CALS. A memo from Michael F. McGrath, DOD CALS Policy Office, states that the CALS initiative is a

> cooperative DOD and industry strategy for transitioning current paper-intensive design and logistic support processes to a highly automated and integrated mode of operation.

The purpose of CALS is to lower costs to the customer (government). CALS became mandatory in 1990 and when fully implemented will allow the government to receive all their data from customers electronically. CALS will also permit them to access the contractor's data base and be on-line in real time to each contractor's data base as it is being developed. The government will be able to check on and evaluate the progress of the following data:

- Engineering Drawings
- Development Specifications
- Training Data
- Provisioning Data
- Reliability Data
- Maintainability Data
- LSA/LSAR Data
- Technical Manuals

CALS allows transition of most technical documentation from paper to electronically stored data. The cost elimination alone is more than significant. The DOD estimates that over $200,000,000 is spent each year to incorporate changes and revisions. Another authority estimates that if the paper documentation (manuals, drawings, etc.) carried on a US Navy fast frigate were eliminated, the ship would float two feet higher in the water. He stated the weight of paper exceeded the weight of ammunition on the ships. Implementation of CALS continues. There are still obstacles and unanswered questions concerning the new system. Decisions are still being made on what is to be entered, how it is to be entered, and how it is to be transmitted and accessed for use. The commercial market offers a large variety of computers and peripherals as well as software. From all these, the government has yet to choose the baseline hardware and software on which to store and access the tremendous amount of data they receive daily. The question of compatibility of contractor hardware and software with the government's baseline must be resolved. The linkup of contractor and government systems must be established. And, for technical manuals, the method for standardizing the format for delivery of data must be defined. Current plans are for text to be prepared with an attached format coding called Standardized General Markup Language (SGML) and for art to be prepared using Initial Graphics Exchange Specification (IGES) and/or Computer Graphics Metafile (CGM).

CALS is to be implemented in two phases. Phase I concentrates on developing a unified digital interface between the government and contractors. It involves development of standards and specifications and preparation of certain technical data products such as engineering drawings and technical manuals. Standards and specifications being developed include the following:

Number	Title
MIL-HDBK-59A	CALS Implementation Guide
MIL-D-28000	Digital Representation for Communication of Product Data: IGES Application Subsets
MIL-M-28001	Markup Requirements and Generic Style Specification for Electronic Printed Output and Exchange of Text
MIL-R-28002	Requirements for Raster Graphics Representation in Binary Format
MIL-D-28003	Digital Representation for Communication of Illustration Data: CGM Application Profile
MIL-STD-1840A	Automated Interchange of Technical Information
FIPS PUB 146	Government Open System Interconnection Profile (GOSIP)

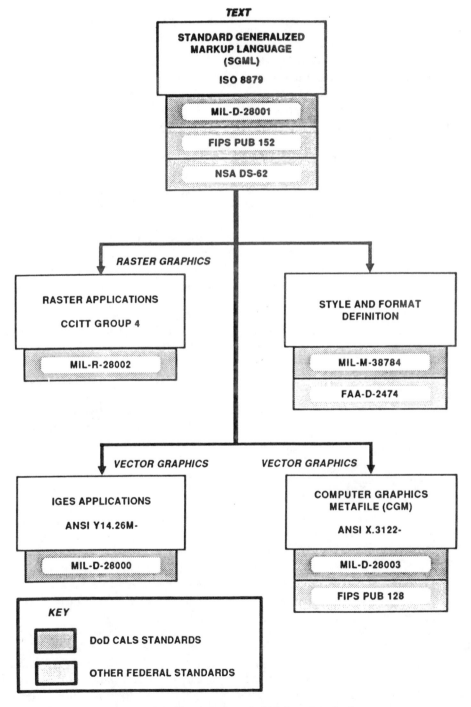

Figure 19-1 *CALS Technical Publishing Standards*

Phase II of CALS will expand on the work of Phase I. Its goal is to complete the evolution of integrated systems between government and contractor. All data which falls under CALS will be transferable between contractor and government, storable in the government's data base, and recallable for editing and modification.

The availability of computers and multiple word processing applications gives the writers a large selection of fonts (type face and point size) to pick from. The contract may define specific requirements, may give a choice of several selections, or may give full freedom to pick and choose. Writers not familiar with the needed hardware or with the particular software may need time for a learning curve. This time must be compensated for in the overall schedule for production of technical publications. It may be necessary to acquire new hardware or software. Or, as a last option, writers may decide to write the manual text and prepare the art on existing hardware and then go outside the company for conversion of the data base to that which meets CALS requirements.

The government is serious about requiring contractors to prepare data on computers which can meet the CALS requirements, and contractors must comply to be competitive. In Phase I of CALS, Government and industry realized that engineering and business data must be integrated with other CALS information. New writers need to understand writing, engineering, and business to operate in the evolving CALS environment.

CALS Data Flow

CALS will involve massive amounts of data flowing between the government and contractors and will require equally large hardware configurations supported by fast and complex software. Figure 19-1 illustrates the basic concept of the CALS data flow. The cycle starts with one or more of the government agencies identifying user requirements. These are passed on to a contractor through a solicitation and contract by electronic communication (wide area network, satellite link, etc.). The contractor builds the required hardware and develops the associated documentation. Subcontractors and vendors may or may not be involved. If they are, either they will have to meet the same requirements for CALS as the prime contractor, or the contractor will bear responsibility for conversion of the vendor/subcontractor data to meet the CALS requirements. The documentation can be in any textual or illustrative form, but whatever is required in accordance with contract requirements will also have to be convertible to CALS. The CALS specifications and standards identify the means by which the data is to be transferred to the government in place of printed manuals and other types of hard copy documentation. Once received and stored in the Government's data base, that data will be available for reuse and updating.

SGML Tagging

The purpose of SGML tagging is to enable transmittal and storage of text from numerous sources and in different formats into a unified, common data base. Data can later be retrieved and restructured into its original format. SGML specifications establish a common markup tag set and page description language. This permits uniformity of documents stored in the database. SGML uses MIL-M-38784 as a tag baseline for the general structure of technical manuals. Tagging refers to the processing of text by the addition of delimiter elements. For example, the following are tags identified as part of the DOD Basic Tag Set in Appendix

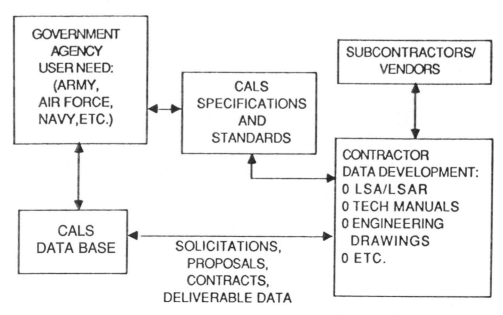

Figure 19-2 CALS Government-Contractor Data Flow

B of the Draft DOD-M-SGML Markup Requirements and Generic Style Specification for Electronic Printed Output and Exchange dated 25 March 1987:

Tag	Definition
⟨nsn security = x⟩	Identifies a National Stock Number. Element start tag is required. End tag may be omitted as appropriate. Security: ''UC'' - unclassified; ''FOUO'' - for official use only; ''NORFON'' - no foreign; ''NNPI'' - naval nuclear propulsion information; ''P'' - proprietary; ''R'' - restricted ''C'' - confidential; ''S'' - secret; ''TS'' - top secret. Default is ''uc.''
⟨paranum security = x⟩	Identifies document paragraph number. Element start tag is required. End tag may be omitted as appropriate. Security: (same as for nsn security).

The following are examples of text with SGML tags:

Table of Contents:

Example 1 ⟨sparar⟩⟨chapnum⟩3⟨sectnum⟩3⟨paranum⟩2⟨title⟩ Maintenance⟨pgno⟩ 3-1

Example 2 ⟨sparar⟩⟨chapnum⟩3⟨sectnum⟩3⟨paranum⟩3⟨title⟩ Lubrication⟨pgno⟩3-2

Text:

Example 3 ⟨paratext⟩Proceed as follows for secondary replacement of insert tube.

Example 4 ⟨subparal⟩It is not necessary to remove the chassis unless resistor R11 must be replaced.

The entire concept of CALS and its development are in the initial stages at this time. As technical writers, you will have to be aware of new developments and procure copies of the specifications and standards as they are released. The technical documentation forms are moving from rigid paperbound to paperless and even to pageless formats.

Illustrations and IGES/CGM Files

Illustrations used by technical writers in their manuals are derived from several sources. In today's electronic world, companies use Computer-Aided Design (CAD), Computer-Aided Manufacturing (CAM), Computer-Aided Engineering (CAE) or some other form of data base system for generating engineering drawings. As the technical publications organizations of these same companies expand into data-based desktop publications, their ability to access the engineering data base increases. Writers can use assembly diagrams, schematic, wiring, cabling, and other forms of engineering drawings to create their own manual-oriented illustrations. Similarly, writers can access illustrations on maintainability and engineering that are already stored in the computer data base. Reusing existing data will lower the cost of generating a ''new'' piece of art. When added together, the number of illustrations derived from this method can have a direct impact on the cost of preparing the manual and can, therefore, lower the overall proposal cost, thus making the company more competitive in the marketplace.

CALS will require delivery of illustrations as well as text by electronic media. Either IGES or CGM will be used for this purpose. MIL-D-28000 defines the IGES application-specific subsets. IGES is based on neutral subset format to allow translation of data between various computers. There are currently four applications subsets in MIL-D-28000:

Class I -   Technical Illustrations

Class II -   Engineering Drawings

Class III -   Electrical/Electronic Applications

Class IV -   Numerical Control Manufacturing

CGM can be used for two-dimensional illustration descriptions by vectors (line segments).

Conclusion

CALS is the wave of the future for all government technical publications. As a result, technical writers must become familiar with the developments of this new approach to paperless technical manuals. Technical manuals will require text and art to be appropriately encoded for electronic transmission to the government's data base in accordance with the various CALS specifications. To grow in the technical writing business, writers will need to be familiar with engineering data and business skills as well as writing skills. Writers must become comfortable using terminals and WYSIWYG (what you see is what you get) screen displays rather than paper.

APPENDIX A

GLOSSARY

AAL	Approved Authorization List
ACO	Administrative Contracting Officer. Government representative whose activities are related to specific contractors. This person is responsible for contract cost and schedule modifications and payments.
ADAD	After Date Award Document (same as ARO)
ADP	Automatic Data Processing
AFAD	Armed Forces Acquisition Document
AFLC	Armed Forces Logistics Command
AFSC	Armed Forces Specialty Code
AFTO	Air Force Technical Order
AMSDL	Acquisition Management Systems and Data Requirements Control List
ANSI	American National Standards Institute. A joint civilian/military organization that develops standards that can be universally used by all DOD agencies and nonmilitary organizations.
Apron	Blank page area at bound edge of a foldout illustration, used primarily for foldout art. By having the first 8-½ inches of a page blank, the art on the remainder of the page can be seen in its entirety while the manual is opened to some other page of text.
AR	Army Regulation. A prefix used for Army publications.
A/R	As required. Used on CDRL's to note that deliverables are due to the customer on an as required basis.
ARO	After Receipt of Order. Contractual date used as initial milestone event. Identifies date contract is awarded.
Art Control Number	A number assigned to each piece of art for identification purposes.
ASCII	American Standard Code for Information Interchange
ASPR	Armed Services Procurement Regulation. A series of regulations that define methods and procedures to be followed when procuring items for the government (replaced by DAR).
ASREQ	As Required.
Assembly	A collection of individual parts or subassemblies which together form a unique item which is usually capable of being disassembled.
ATE	Automatic Test Equipment
ATP	Acceptance Test Procedure. A test procedure prepared by engineering. It defines and quantitizes all parameters of an item which must be met for sell-off of that item.
AVUM	Aviation Unit Maintenance
B&P	Bid and Proposal. Usually refers to funds available for bid and proposal activity.
Bid Package	A collection of technical data defining the content (hardware) and function of an item, compiled specifically to provide a basis upon which bidders may perform a cost analysis. The package is normally made available to bidders during a bidders conference.
BII	Basic Issue Item
BI-MO	Bi-monthly (delivery every two months)
BI-WE	Bi-weekly (delivery every two weeks)
BIT	Built-In-Test. A test built into an item, which, when activated, will test the item to some predetermined extent to ascertain the ability of the item to perform its design function, and, in some cases, to fault isolate should the item be defective.
Black Box	Term used to identify an item whose content and functioning internal parts are not described in a technical manual. All discussions are centered around its inputs and outputs.
Block Diagram	Diagram representing hardware, functions, or combination of both, depicted by blocks rather than discrete parts. May be either simplified or detailed in complexity.

BOA	Basic Ordering Agreement
BOE	Basis of Estimate. A cost-related estimate identifying how individual costs cited in a contract were derived. Cost estimates may be related to capital equipment, labor, various judgmental factors, engineering estimates, and contingencies. The need for BOE data is based on the Truth in Negotiation Act, Public Law 87-653, and Federal Procurement Regulation. The Department of Defense Contract Pricing Proposal (DD Form 633-2), "Instructions to Offerors" states, "As part of the specific information required by this form, the offeror must submit with this form, and clearly identify as such, cost or pricing data (that is, data which is verifiable and factual and otherwise as defined in ASPR 3-807.3 (e)."
C/A	Cable Assembly
CAD	Computer-Aided Design
CAE	Computer-Aided Engineering
CAGE	Commercial and Government Entity Code
CALS	Computer-Aided Acquisition and Logistics Support
CAM	Computer-Aided Manufacturing
CCB	Configuration Control Board
CDR	Critical Design Review. That point in a program (a major milestone) where detail design has been completed and no further major changes are anticipated. At this point, data is usually sufficient to start the technical manual writing effort.
CDRL	Contract Data Requirements List. (DD Form 1423). List of data items to be delivered under a government contract.
CECOM/SATCOMA	Communications Electronics Command/ Satellite Communications Agency
CFAE	Contractor-Furnished Aerospace Equipment
CFE	Contractor-Furnished Equipment
CGM	Computer Graphics Metafile
CM	Configuration Management
CM	Corrective Maintenance
COEI	Components of End Item
Concurrent Delivery	Delivery of both an end item and support items at the same time.
Contract Pricing Proposal	DOD Form DD 633-2. It is used by prospective bidders (offerors) to identify summary proposal costs.
CONUS	Continental United States
Corrective Maintenance	Unscheduled maintenance performed on an item for purpose of returning that item to its normal operational condition.
COTS	Commercial Off-the-Shelf.
CPFF	Cost Plus Fixed Fee. A type of contract.
CPIF	Cost Plus Incentive Fee. A type of contract.

CRC	Camera Ready Copy
CTC	Cost to Complete
CTD	Cost to Date
Cut-away	Area on an assembly drawing which indicates part of an assembly is "cut" out to permit a view of what would otherwise be hidden, sometimes called a rat hole.
DAC	Days After Contract
DAR	Defense Acquisition Regulations. Government publication identifying regulations relating to acquisitions of government agencies (replaced by FAR, Federal Acquisition Regulations).
DCAA	Defense Contract Audit Agency. A government agency concerned with the government's ability and need to monitor and control the relationship between parties to contracts let by government agencies.
DCAS	Defense Contract Administration Service
DCN	Drawing Change Notice
DD-250	Material Inspection and Receiving Report. A government form by which the customer accepts and acknowledges receipt of contractual items. It is also the means by which payment is made by the customer to the contractor.
Depot Maintenance	Highest level of maintenance. Includes overhaul, rebuild, and calibration.
DID	Data Item Description (DD Form 1664). Identifies specific contract deliverable data requirements.
DIP	Dual In-line Package
DL	Direct Labor. Labor directly attributable to a job involved in development of the program deliverables: includes engineering, manufacturing, and support efforts. DL serves as a basis for overhead.
DM	Data Manager/Management
DMWR	Depot Maintenance Work Request
DOD	Department of Defense. The constituent part of government relating to the military departments (Army, Navy, Air Force, and Marines).
DODISS	Department of Defense Index of Specifications and Standards
DS	Direct Support (Maintenance)
DT&E	Development Test and Evaluation
DT/OT	Design Test/Operational Test
DTUPC	Design to Unit Production Cost
ECN	Engineering Change Notice
ECO	Engineering Change Order
ECP	Engineering Change Proposal
ED	Equipment Delivery (date of)
EIA	Electronic Industries Association
EIR	Equipment Improvement Recommendation

End Item	Final (top assembly) item of deliverable hardware. The item may consist of assemblies, subassemblies, and component parts.
Equipment Tree	A block structure drawing which shows the hierarchy breakdown of a piece of equipment, system, etc.
ES&ML	Expendible/Durable Supplies and Materials List
Exploded View	Isometric illustration depicting all piece parts of an end item or any of its assemblies or subassemblies. The drawing may contain flow lines (used to depict the sequence of assembly) and callouts for parts identification.
Family Tree	Diagrammatic representation of highest to lowest order of relationships between all parts of an item. These diagrams are laid out such that the highest (top) assembly is listed at the top of the page, and each successive lower assembly is listed below, in descending order.
FAR	Federal Acquisition Regulations (replaces DAR)
FCA	Functional Configuration Audit
FFP	Firm Fixed Price; a type of contract
FGC	Functional Group Code
Field Maintenance	The category of maintenance activities done at echelons below depot maintenance; may include operator, organizational, shop, intermediate, or direct support levels.
FIPS	Federal Information Processing Standard
FOMM	Function-Oriented Maintenance Manual
Font	A complete assortment of type in one size and style.
Formal Advertising	Related to fixed-price contracts let by the government. Used to initiate competitive bidding between prospective contractors for goods and/or services.
Format	Particular mode of usage, arrangement, content of a technical manual (e.g., MIL-M-38784 is a format specification and MIL-M-38798 is a content specification.)
FPI	Fixed Price Incentive
FQT	Functional Qualification Test
FSCM	Federal Supply Code for Manufacturers
FSD	Full Scale Development
FSN	Federal Stock Number
Full Page	Typically an 8-½ × 11-inch page, containing text, art, or tabular material within the total image area (e.g. 7 × 9-inch area on the page). Running heads and feet are outside the image area.
Functional Diagram	Illustration depicting an item or part of an item in a functional (as opposed to a hardware) format. Functional diagrams are usually presented with a left-to-right flow, such that the beginning (inputs) are on the left, and end (outputs) are on the right. Symbols used to represent piece parts are dependent upon the type of diagram being illustrated, and the specifications cited by the contract or customer.
G&A	General and Administrative. Costs related to management and business which are not directly identifiable with individual program costs, but are required for the program to be performed. Capital equipment and facilities are examples.
Gantt Chart	Bargraph representation of relationships between events or characteristics. Typically, it depicts the relationship between program events and time.
GAO	General Accounting Office
GFE	Government Furnished Equipment
GFI	Government Furnished Information
GFP	Government Furnished Property
GOSIP	Government Open Systems Interconnection Profile
GPO	Government Printing Office. A central government facility which acts as a central point for storage and printing of all government documents.
GS	General Support (Maintenance)
GSA	General Services Administration
GSERD	Ground Support Equipment Recommendation Data
H/W	Hardware
IAW	In Accordance With
IDL	Indirect Labor
IEEE	Institute of Electrical and Electronics Engineers
IFB	Invitation For Bid
IGES	Initial Graphics Exchange Specification
ILS	Integrated Logistics Support
IPB	Illustrated Parts Breakdown
IPR	In-Process Review
IR&D	Independent Research and Development or Internal Research and Development
ISO	International Organization for Standardization
JETDS	Joint Electronics Type Designation System
K/O	Kick-Off. Usually in reference to a conference; the first conference convened to discuss contract performance.
LAN	Local Area Network
Leading	The amount of space between lines, derived from the boarders of a piece of lead type.
Light Box	A box having a translucent glass top and containing one or more bulbs, used to strip (move) information from one page into another by cutting and pasting.

Long Lead Items	Items which require a long time to procure due to their limited availability (caused by factors such as complexity in design or manufacture) or which have limited time available for procurement.	OPEVAL	Operational Evaluation
		OTIME	One Time (CDRL delivery)
		OTS	Off-the-shels
		PCA	Physical Configuration Audit
LRIP	Low Rate of Initial Production	PCB	Printed Circuit Board
LRU	Lowest (Line) Replaceable Unit	PCO	Procuring Contracting Officer. A government representative who evaluates requirements, prepares RFP's and RFO's for contractors to bid, and evaluates the contractors' proposals.
LSA	Logistics Support Analysis		
LSAR	Logistics Support Analysis Record		
LT	Letter of Transmittal		
MAC	Maintenance Allocation Chart		
MAC	Months After Contract	PDES	Product Data Exchange Specification
Maintenance Concept	Statement (usually in the logistics section of the design specification) of what maintenance is to be done, by whom, how often, with what resources, and at what levels. It is usually defined by the customer in the feasibility or conceptual design phase.	PDL	Page Description Language
		PDR	Preliminary Design Review
		PERT	Program Evaluation and Review Technique
		Pica	Printing measure of linear (horizontal or vertical) distance; 1 inch = 6 picas; 1 pica = 12 points.
Manual Tree	A block structure drawing showing the hierarchy of manuals which describe/support a piece of equipment.	PIL	Preferred Items List
		P/L	Parts List
		PM	Preventive Maintenance
MAPS	Maintenance Action Precise Symptom	PM	Program Management
MEAR	Maintenance Engineering Analysis Record	PMCS	Preventive Maintenance Checks and Services
Milestone Chart	A chart used to depict major and minor events of a program from start to finish. It usually depicts key contractual events and hardware related events (such as PDR and CDR).	PMI	Preventive Maintenance Instructions
		PMO	Program Management Office
		PMWC	Preventive Maintenance Work Cards
		P/N	Part Number
		PO	Purchase Order
MIL-STD	Military Standard	Point	0.0138 inch (1/72): used to specify type size.
MOS	Military Occupation Specialty		
MTBF	Mean Time Between Failures	PPL	Provisioning Parts List
MTHLY	Monthly (related to CDRL delivery)	PQT	Physical Qualification Test
MTTR	Mean Time To Repair	Provisioning	The process of acquiring parts used to repair purchased hardware.
New Look	A United States Army Technical Manual format employing simplified content and layout of technical data, identified in MIL-M-63036.		
		PTO	Preliminary Technical Order
		PWB	Printed Wiring Board
NLT	Not Later Than	QA	Quality Assurance
NSN	National Stock Number. A government inventory stock number (13 digits long) used for material management.	QC	Quality Control
		QQPRI	Qualitative and Quantitative Personnel Requirement Information
OCONUS	Outside Continental United States	QRTLY	Quarteriy (CDRL delivery related)
ODC	Other Direct Costs. Costs related to performance of a contract which have a specific objective such as printing and making negatives. Material, travel, and DL are not considered part of ODC. These are costs incurred which are not directly attributable to a unique program.	R&D	Research and Development
		RCM	Reliability Centered Maintenance
		Repro	Reproducible art and text pages used as originals for reproduction purposes.
		RFP	Request for Proposal. A request from the government for a proposal input by the contractor. The RFP contains pertinent information required by the customer to prepare the technical and cost inputs.
OH	Overhead (costs). Costs indirectly related to contract performance. Examples are accounting and facilities.		
		RFQ	Request for Quote. Performs the same function as the RFP.
OJT	On-The-Job Training		
O&M	Operation and Maintenance	R&M	Reliability and Maintainability
ONE/R	One time with revisions (related to CDRL delivery)	ROM	Reasonable Order of Magnitude (in reference to costing)

Rough Draft	Manuscript (first cut) used for validation, first delivery, In-Process Reviews, etc. It may or may not be complete in terms of format or content.
RPSTL	Repair Parts and Special Tools List
Rubylith	A clear mylar base film coated with a red plastic emulsion. The emulsion (which can be cut and peeled) acts as a filter for photographic processing.
Running Heads and Feet	Information placed in the upper and lower part of a page, such as a volume number, page number, chapter title, etc. This information is outside the "image area."
SGML	Standardized General Markup Language
SOW	Statement of Work. Customer's statement of requirements for performance on a contract; normally it is made a part of the contract.
SMR	Source, Maintenance and Recoverability
TCP/IP	Transmission Control Protocol/Internet Protocol
TMSS	Technical Manual Specifications and Standards

TO	Technical Order
TOPP	Technical Order Publications Plan
TRADOC	Training and Doctrine Command
Tree	A type of block drawing which shows a structure or hierarchy of items; examples include a drawing tree, equipment tree, and a manual tree.
Turn Page	Full page (typically $8\frac{1}{2} \times 11$-inches) with data rotated 90 degrees clockwise. To read a turn page, you rotate the book so its right side becomes the bottom.
Val	Validation
VCR	Video Cassette Recorder
Ver	Verification
WBS	Work Breakdown Structure. A control document designed to structure program elements in a hierarchial manner. The matrix of vertically and horizontally aligned boxes permits organized numbering of functional elements and tasks for management control purposes such as cost tracking and control.
W/L	Wire List
WUCM	Work Unit Code Manual

INDEX